Lecture Notes in Mathematics

Edited by A. Dold and B. Eckmann

643

Michael Davis

Multiaxial Actions on Manifolds

Springer-Verlag
Berlin Heidelberg New York 1978

Author
Michael Davis
Mathematics Department
Columbia University
New York, NY 10027/U.S.A.

Library of Congress Cataloging in Publication Data

Davis, Michael, 1949-
Multiaxial actions on manifolds.

(Lecture notes in mathematics ; 643)
Includes bibliographical references and index.
1. Topological transformation groups.
2. Lie groups. 3. Manifolds (Mathematics)
I. Title. II. Series: Lecture notes in mathematics (Berlin) ; 643.
QA3.L28 no. 643 [QA613.7] 510'.8s [512'.55]
78-3765

AMS Subject Classifications (1970): 57E15

ISBN 3-540-08667-6 Springer-Verlag Berlin Heidelberg New York
ISBN 0-387-08667-6 Springer-Verlag New York Heidelberg Berlin

Printing and binding: Beltz Offsetdruck, Hemsbach/Bergstr.
2141/3140-543210

PREFACE

These are the notes for a series of five lectures which I gave in the Transformation Groups Seminar at the Institute for Advanced Study during February of 1977. They concern the study of smooth actions of the compact classical groups (O(n), U(n) or Sp(n)) which resemble or are "modeled on" the linear representation $k\rho_n$. In the literature such actions have generally been called "regular" O(n), U(n) or Sp(n)-actions; however, following a suggestion of Bredon, I have adopted the terminology "k-axial actions."

My interest in these actions was ignited by the beautiful theory of biaxial actions on homotopy spheres discovered by the Hsiangs, Jänich, and Bredon. Perhaps the most striking result in area is the theorem, due to the Hsiangs and independently to Jänich, which essentially identifies the study of biaxial O(n)-actions on homotopy spheres (with fixed points) with knot theory. Also of interest is Hirzebruch's observation, that many Brieskorn varieties support canonical biaxial actions. Such material is discussed in Chapter I.

In my thesis, I studied the theory of k-axial O(n), U(n) and Sp(n) actions for arbitrary k such that $n \geq k$. The main result, here called the Structure Theorem, is proved (in outline) in Chapter IV. This theorem implies that (assuming a certain obviously necessary condition) any k-axial action is a pullback of its linear model. In Chapter VI, I indicate how this result can be combined with Smith theory and surgery theory in order to classify all such actions on homotopy spheres up to concordance.

Many proofs are omitted and some are only sketched. In the case of U(n) or Sp(n)-actions the classification up to concordance is joint work with W. C. Hsiang. In the case of O(n)-actions it is joint work with Hsiang and J. Morgan.

I have added two appendices to the original version of these notes. The first one deals with the theories of biaxial actions of SO(3), G_2 or SU(3), which are surprisingly different from the corresponding theories of biaxial actions of O(3), SO(7) or U(3). The second appendix contains the proof of an important technical lemma which was stated in Section A.2 of Chapter V, but not proved there. This lemma deals with the homological relationships between various pieces of the orbit spaces of two regular O(n)-manifolds which have the same homology.

My thanks go to Bill Browder, Glen Bredon, Wu-yi Hsiang, John Morgan and Gerry Schwarz for many helpful conversations. I would also like to thank Deane Montgomery and the other participants in the seminar for their interest and encouragement. I am particularly indebted to my thesis advisor, Wu-chung Hsiang, not only for his help in that venture, but also for the many continuing discussions I have had with him during the past three years. I was supported by NSF grant MCS72-05055 A04 while I was at the Institute.

Columbia University
November, 1977

CONTENTS

0. INTRODUCTION

Suppose that G is a compact Lie group and that Σ^m is a homotopy m-sphere. We wish to consider smooth actions $G \times \Sigma \to \Sigma$ of G on Σ. One reason for studying such objects is that there are natural examples; namely, if $G \hookrightarrow O(m+1)$ is any representation, then G acts on the unit sphere in R^{m+1}. The action $G \times S^m \to S^m$ so obtained is called a "linear action."

One standard approach to this problem is to

(1) Make some innocent sounding hypothesis (e.g. there is an orbit of codimension one or two, the principal orbit type is a Stiefel manifold, or that G is simple and $\dim \Sigma < \dim G$) to insure that the orbit types of G on Σ will be the same as those of some linear action, and then

(2) try to classify all those smooth actions with the same orbit types as some given linear action.

This approach was first explicitly formulated by the Hsiangs. Problems (1) and (2) are closely related and often solved by the same person (or persons) in a single paper; however, we shall concentrate on (2).

If one wants to consider a relatively large group G, then the first representation one might try in the above program is the standard representation of $O(n)$ on R^n. Historically, this was the representation that was first considered. Next one might consider the standard representation of $U(n)$ on $\mathbb{C}^n$ or $Sp(n)$ on $\mathbb{H}^n$ ($\mathbb{H}$ = quaternions). We will use the notation

$$\rho_n: G^d(n) \hookrightarrow O(dn)$$

to denote one of these representations, where $G^d(n) = O(n)$, $U(n)$ or $Sp(n)$ as $d = 1$, 2 or 4. Finally one might consider other representations such as the adjoint representation or multiples of the standard representation. In these lectures I will deal only with smooth actions which are modeled on multiples of the standard representation (i.e., "modeled on $k\rho_n$").

Definitions: Suppose that G acts smoothly on M. The slice representation at $x \in M$ is the action of G_x on the subspace of T_xM which is perpendicular to the orbit, i.e., on

$$S_x = T_xM/T_x(G(x))$$

where G_x is the isotropy group and $G(x)$ is the orbit passing through x. Let F_x be the subspace of S_x left pointwise fixed by G_x and let

$$V_x = S_x/F_x.$$

The G_x-module V_x is called the normal representation at x. There is an equivalence relation on pairs of the form (G_x, V_x) induced by conjugation on the subgroup and equivariant linear isomorphisms on the vector space. The equivalence class of (G_x, V_x) is called the normal orbit type of x.

Let W be any G-module. We shall say that M is modeled on W if the normal orbit types of G on M occur among the normal orbit types of W. M is a k-axial $G^d(n)$-manifold if it is modeled on $k\rho_n$. The following material concerns such actions.

I. SOME HISTORY

Bredon has written a survey article [4], largely devoted to monoaxial and biaxial actions on homotopy spheres. The interested reader is also referred to [5], [12] and [14]. Let us note that [4] and [5] contain excellent bibliographies.

A. Actions modeled on $1\rho_n$.

1. The example of Montgomery and Samelson: In 1961 in [19], Montgomery and Samelson considered the following construction. Let B^m be a smooth contractible manifold with boundary. Let

$$X = D^n \times B^m.$$

We can let $O(n)$ act on X via the standard representation on D^n and trivially on B^m. Let us make a few observations.

1) The action on X is modeled on $1\rho_n$.

2) ∂X is simply connected.

3) So, by the h-cobordism Theorem, X is a disk (except possibly when $m = 3$ and $n = 1$).

4) $D^n/O(n) \cong [0,1]$ and, therefore, $X/O(n) \cong [0,1] \times B^m$.

5) Since $\partial X = S^{n-1} \times B^m \cup D^n \times \partial B^m$

$\partial X/O(n) \cong B^m \cup [0, 1] \times \partial B^m \cong B^m$.

6) The fixed point set of $O(n)$ on ∂X is diffeomorphic to ∂B^m (which can be an arbitrary integral homology sphere bounding a contractible manifold).

7) Thus, there are infinitely many inequivalent monoaxial $O(n)$-actions on S^{n+m-1} which are distinguished from one another by their orbit spaces (or by the fundamental groups of their

fixed points sets).

The same construction works for U(n) or Sp(n) by replacing D^n by D^{2n} or D^{4n} (the unit disk is the standard representation).

These examples are not really so different from the linear action $\rho_n + m1$ on S^{n+m-1} (where $B^m = D^m$). In fact, we shall see that this type of phenomenon of replacing the linear orbit space with a space whose strata have the correct homology, can always happen. The action on ∂X is "concordant to the linear action" (since it extends to an action on X which is a disk). To see this, remove a small disk about a fixed point in X to obtain an action on $S^{n+m-1} \times [0,1]$ which is the original action on ∂X on one end and equivalent to the linear action on the other end.

2. Wu-yi Hsiang's thesis: In his thesis which appeared as [16], Wu-yi Hsiang proved that every effective, smooth SO(n)-action (or O(n)-action) on a π-manifold of sufficiently small dimension (compared to n) was monoaxial. In a related joint paper with W. C. Hsiang [13], it was proved that a smooth $G^d(n)$-action on a homotopy sphere was monoaxial provided the principal orbits were spheres (of dimension (dn-1)). Similar "regularity" theorems had been proved earlier by Bredon [3], Wang, Montgomery, Samelson and Yang (see [4], [9] and [17] for further discussion).

Wu-yi Hsiang also showed that the examples of Montgomery and Samelson were the only possibilities of monoaxial actions on homotopy spheres (and hence that no exotic sphere admits a monoaxial $G^d(n)$-action).

<u>Theorem 1</u> (W. Y. Hsiang): <u>There is a bijection (assuming</u> $d \neq 1$ <u>and</u> $n \neq 1$

$$\left\{\begin{array}{l}\text{equivariant diffeomorphism}\\ \text{classes of monoaxial}\\ G^d(n)\text{-actions on homotopy}\\ \text{spheres of dimension}\\ dn+m-1\end{array}\right\} \longleftrightarrow \left\{\begin{array}{l}\text{diffeomorphism}\\ \text{classes of}\\ \text{contractible}\\ \text{manifolds with}\\ \text{boundary}\end{array}\right\}$$

More precisely, Hsiang showed that if $G^d(n)$ acts monoaxially on Σ^{dn+m-1}, then

(1) $\Sigma/G^d(n) = B^m$ is a contractible manifold with boundary.

(2) Σ and Σ' are equivalent if and only if their orbit spaces B and B' are diffeomorphic.

Let us now prove statements (1) and (2). The ideas of the proof will occur again in our considerations of general multiaxial actions.

<u>Proof of (1)</u>: First, why is B a manifold with boundary? (It is certainly not generally true that the orbit space of an arbitrary smooth G-manifold is a manifold.) In our case, if x is a fixed point, then it follows from the Differentiable Slice Theorem that x has an invariant neighborhood in Σ of the form $R^{dn} + R^{m-1}$ where $G^d(n)$ acts via $1\rho_n$ on R^{dn} and trivially on R^{m-1}. Since $(R^{dn} + R^{m-1})/G^d(n) \cong [0,\infty) \times R^{m-1}$, B is a manifold with boundary near $\pi(x) \in B$. If x is not a fixed point, then it again follows from the Slice Theorem, that x has an invariant neighborhood of the form $S^{dn-1} \times R^m$; and hence, $\pi(x)$ has a neighborhood isomorphic to R^m. □

Now to finish the proof that B is contractible we show that

(a) B is simply connected.

(b) B is acyclic.

First we need two lemmas.

Lemma 2: If $d = 2$ or 4, then the fixed point set $F = \Sigma^{G^d(n)}$ is an integral homology $(m-1)$-sphere. In the $O(n)$ case, if n is even then F is also an integral homology $(m-1)$-sphere.

Proof: Let T be a maximal torus of $G^d(n)$. The lemma will follow from P. A. Smith's Theorem if we can show that $F = \Sigma^T$. It suffices to check this on each orbit and since each orbit occurs as an orbit in the linear model it suffices to check it there. But clearly $F(G^d(n), R^{dn}) = \{0\} = F(T, R^{dn})$, except in the case of $O(n)$ when n is odd (in which case $F(T, R^n) = F(O(n-1), R^n) \cong R^1$. □

Lemma 3: Suppose G acts smoothly on M with precisely one orbit type, say, G/H. Then $G/H \to M \to M/G$ is a smooth fiber bundle with structure group N_H/H. Moreover, the associated principal bundle can be identified with $N_H/H \to M^H \to M^H/N_H = M/G$.

Proof: This is a well-knwon and important lemma (see, for example, [5]). That $M \to M/G$ is a smooth fiber bundle follows from the Slice Theorem. The structure group is the group of equivariant diffeomorphisms of G/H which is easily seen to be N_H/H (N_H is the normalizer of H in G). The last sentence is also an easy exercise. □

Proof of (a): Since B is a manifold with boundary, to show that B is 1-connected it suffices to show its interior int B is 1-connected. By Lemma 2, we have a fiber bundle $S^{dn-1} \to \Sigma - F \to \text{int } B$. Except in the case $G^d(n) = O(1)$, S^{dn-1} is connected so $\pi_1(\Sigma - F) \to \pi_1(\text{int } B)$ is an epimorphism. But we have the commutative diagram

$$\begin{array}{ccc} \pi_1(\Sigma - F) & \longrightarrow & \pi_1(\text{int } B) = \pi_1(B) \\ & \searrow \quad \swarrow & \\ & \pi_1(\Sigma) & \\ & \| & \\ & 0 & \end{array}$$

It follows that $\pi_1 B() = 0$.

$\square$

<u>Proof of (b)</u> (that B is acyclic): We will use the symbols $X \underset{Z}{\sim} Y$ to mean that X and Y have the same integral homology. Now let

$$X = \Sigma\text{-tubular nbhd. of } F$$

$$E_1 = X^{G(n-1)}$$

$$\bar{B} = X/G(n) \cong B\text{-collar of } \partial B$$

(F is the fixed point set). By Lemma 2, E_1 is a principal $G^d(1)$-bundle over $\bar{B}$ ($G^d(1) = N_{G^d(n-1)}/G^d(n-1)$). In the U(n) and Sp(n) cases we use Lemma 1 to show that

$$X \underset{Z}{\sim} S^{dn-1}$$

and therefore

$$E_1 \underset{Z}{\sim} S^{d-1}.$$

So it follows from a simple spectral sequence argument that $\bar{B} \underset{Z}{\sim} pt$. In the O(n) case we argue as follows. First E_1 is the trivial O(1) bundle since $\bar{B}$ is simply connected. Since X is an associated bundle

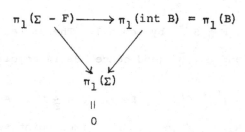

$$X = S^{n-1} \times \bar{B} .$$

Now, we have 2 cases:

<u>n is even</u>: First, $F \underset{Z}{\sim} S^{m-1}$ by Lemma 1. Therefore, $X \underset{Z}{\sim} S^{n-1}$ by Alexander Duality and so, $\bar{B}$ (and hence B) is acyclic.

<u>n is odd</u>: Let $A = \Sigma^{O(n-1)}$. By Lemma 1, $A \underset{Z}{\sim} S^m$. A is a $O(1)$-manifold where $O(1) = N_{O(n-1)}/O(n-1)$. The fixed point set of $O(1)$ on A is F (which is of codimension one in A) and $A/O(1) \cong B$. In [20], Yang proved that if $O(1)$ acts on an integral homology sphere with fixed point set of codimension one, then the orbit space is acyclic. Here is an argument. Consider the exact sequence of the pair (A,F). For $i \neq 0$ or m, we have $H_i(A,F) \cong H_{i-1}(F)$. Since $O(1)$ acts trivially on F, it acts trivially on $H_i(A,F)$. On the other hand by excision,

$$\begin{aligned} H_i(A,F) &\cong H_i(E_1, \partial E_1) \\ &\cong H_i((\bar{B}, \partial\bar{B}) \times O(1)) \\ &\cong H_i(\bar{B}, \partial\bar{B}) + H_i(\bar{B}, \partial\bar{B}) . \end{aligned}$$

Since $O(1)$ operates on these last terms by switching factors and since it acts trivially on $H_i(A,F)$, it follows that $H_i(\bar{B}, \partial\bar{B}) = 0$, $i \neq m$. Thus B is acyclic. This completes the proof of (1). Arguments similar to this are due to Borel and Bredon (see [1] and [2]). □

Now, we turn to the proof of (2). Actually, we can prove the following.

<u>Proposition 4</u>: <u>Suppose that</u> M <u>and</u> M' <u>are monoaxial</u> $G^d(n)$ <u>manifolds over</u> B <u>and</u> B'. <u>Suppose further that the bundles of principal</u>

orbits are trivial (this is implied for example by B and B' being acyclic). Then M and M' are equivariantly diffeomorphic if and only if B and B' are diffeomorphic.

Proof: If $M \to M'$ is an equivariant diffeomorphism, then the induced map $B \to B'$ is a diffeomorphism.

In the other direction we would like to say that if $f\colon B \to B'$ is a diffeomorphism, then there exists an equivariant diffeomorphism $\tilde{f}\colon M \to M'$ covering f.

$$\begin{array}{ccc} M & \xrightarrow{\tilde{f}} & M' \\ \downarrow & & \downarrow \\ B & \xrightarrow{f} & B' \end{array} .$$

In fact, this is true; however, proving that $\tilde{f}$ is differentiable near a fixed point is a somewhat subtle point (essentially involving the Covering Isotopy Theorem, see [5]). We proceed a little differently.

Let N be the normal bundle of the fixed point set F in M. First we claim that N is a trivial $G^d(n)$-vector bundle. The fiber of N is R^{dn} and the structure group can be reduced to the centralizer of $1\rho_n$. This centralizer is $GL^d(1)$, the general linear group over $\mathbb{R}$, $\mathbb{C}$, or $\mathbb{H}$. Picking an invariant metric for N we can actually reduce the structure group to $G^d(1)$. Let $E_0 \to F$ be the associated principal bundle. Consider the unit sphere bundle of N

$$\partial N = S^{dn-1} \times_{G(1)} E_0.$$

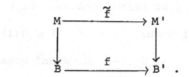

Let $E_1(\partial N) = \partial N^{G(n-1)}$ be the total space of the principal bundle associated to the orbit bundle $\partial N \to \partial N/G(n)$. Then

$$E_1(\partial N) = G(1) \times_{G(1)} E_0$$
$$\cong E_0.$$

Now pick equivariant tubular neighborhoods $\tau : N \to M$ and $\tau' : N' \to M'$ covering collared neighborhoods $\partial B \times [0,1] \to B$ and $\partial B' \times [0,1] \to B'$. Let $\bar{B}$ = B-collared neighborhoods. Then τ induces a bundle isomorphism $E_1(\partial N) \cong E_1(M)|_{\partial\bar{B}}$, where $E_1(M) \to \bar{B}$ is the principal bundle associated to the bundle of principal orbits. Since $\bar{B}$ is contractible, $E_1(M)$ is the trivial bundle, i.e. $E_1(M) \cong G(1) \times \bar{B}$. Thus, $E_1(\partial N)$ and E_0 are also trivial $G(1)$-bundles and consequently

$$N \cong R^{dn} \times F.$$

We may assume that $f: B \to B'$ is a product map on the collared neighborhoods. Pick any equivariant diffeomorphism $f: S^{dn-1} \times \bar{B} \to S^{dn-1} \times \bar{B}'$ covering f.

$$\begin{array}{ccc} S^{dn-1} \times \bar{B} & \xrightarrow{\tilde{f}} & S^{dn-1} \times \bar{B}' \\ \downarrow & & \downarrow \\ \bar{B} & \xrightarrow{f} & \bar{B}' \end{array} .$$

Consider the restriction of $\tilde{f}$ to $\partial N = S^{dn-1} \times \partial\bar{B}$. This is a bundle map and on each fiber it is multiplication by an element of $G^d(1)$. Since the action of $G^d(1)$ on S^{dn-1} is scalar multiplication, $\tilde{f}|_{\partial N}$ extends to a linear map of disk bundles $N \to N'$. Using this extension we get an equivariant diffeomorphism $\tilde{f}: M \to M'$. □

B. Actions modeled on $2\rho_n$.

Things began to heat up with the work of Bredon, the Hsiangs, and Jänich on biaxial actions. We shall first consider $O(n)$-actions and then discuss the $U(n)$ and $Sp(n)$ cases (which are similar).

1. Bredon's examples: These examples are similar to the linear action $2\rho_n: O(n) \times S^{2n-1} \to S^{2n-1}$. This action has two types of orbits. The principal orbits are second Stiefel manifolds $O(n)/O(n-2)$ and the singular orbits are spheres $O(n)/O(n-1)$. The orbit space is a 2-disk. If we consider biaxial actions on homotopy $(2n - 1)$-spheres, then an argument similar to the one which we gave for monoaxial actions shows that the orbit space is a contractible 2-manifold, hence, a 2-disk. However, the other part of our argument does not work, since it is not true that any equivariant diffeomorphism of the normal sphere bundle of the singular orbits extends to the disk-bundle.

Bredon's program was to classify Lie group actions on homotopy $(2n-1)$-spheres with orbit space D^2. In 1965 in [3], he shows that for each nonnegative integer k, there is a biaxial $O(n)$ manifold M_k^{2n-1} with the following properties.

(1) $M_k^{2n-1}/O(n) \cong D^2$.

(2) For n odd and k odd, M_k^{2n-1} is homeomorphic to S^{2n-1}.

(3) M_k^3 is the lens space $L^3(k,1)$.

(4) M_k^{2n-1} is highly connected and for n even, $H_{n-1}(M_k^{2n-1}) \cong Z_k$.

(5) The fixed point set of $O(1)$ on M_k^{2n-1} is equal to M_k^{2n-3}.

(6) These are the only biaxial actions with orbit space D^2.

Remarks: Actually, Bredon only constructed these examples for k odd. Also, in his notation M_k^{2n-1} is what we call M_{2k+1}^{2n-1}.

Of course, it later developed (as was shown by Hirzebruch [12]) that Bredon's examples coincided with certain natural actions on Brieskorn varieties. To be more specific, let

$$f(z) = (z_1)^k + (z_2)^2 + \ldots + (z_{n+1})^2$$

and let

$$\Sigma_k^{2n-1} = f^{-1}(0) \cap S^{2n+1} .$$

The linear action of $O(n)$ operating on the last n coordinates leaves the intersection invariant. It is not hard to check that this action is biaxial with orbit space D^2. Since these actions are classified, they coincide with Bredon's examples and, in fact, $M_k^{2n-1} = \Sigma_k^{2n-1}$.

It was shown by Hirzebruch [12] and independently by the Hsiangs [15] that for n odd, M_k^{2n-1} is the Kervaire sphere if $k \equiv \pm 3$ mod 8 and it is the standard sphere if $k \equiv \pm 1$ mod 8. In particular, this means that M_3^{2n-1} is not concordant to the linear action. In fact, it is not too difficult to see that the integer k is a concordance invariant ($k - 1$ essentially is an index-type invariant).

2. <u>The classification theorem of the Hsiangs and Jänich for actions with two types of orbits</u>: In an important paper [14], the Hsiangs, and independently Jänich in [18], gave a different proof of Bredon's classification theorem. Their proof also generalized to give a classification theorem for certain kinds of actions over a given orbit space with exactly two types of orbits. This classifica-

tion was in terms of a certain invariant called the "twist invariant" by the Hsiangs and the "characteristic reduction" by Jänich.

The Hsiangs applied this result to biaxial U(n) or Sp(n) actions on homotopy (2dn-1)-spheres. If $G^d(n) \times \Sigma^{2dn-1} \to \Sigma^{2dn-1}$ is biaxial, then it follows that the fixed point set is empty and that the orbit space B^{d+1} is a contractible manifold with boundary (of dimension 3 in the U(n) case and of dimension 5 in the Sp(n) case). The biaxial actions over a given orbit space were again classified by the non-negative integers (if B^{d+1} is a disk, then these actions are the restriction to $G^d(n)$ of the O(dn)-action on M_k^{2dn-1}). However, only the action corresponding to the integer 1 is a homotopy sphere. Thus, the Hsiangs showed that for d = 2 or 4 biaxial $G^d(n)$-actions on homotopy (2dn-1)-spheres were in 1 - 1 correspondence with contractible (d+1)-manifolds. Furthermore, these actions are concordant to the linear action.

3. Knot manifolds: In the same two papers [14] and [18], the Hsiangs and Jänich also considered biaxial O(n)-actions with fixed points. Here, the linear action they had in mind was $2\rho_n + (m+1)1$ (two times the standard representation plus the (m+1)-dimensional trivial representation) of O(n) on S^{2n+m}. This has three types of orbits: Stiefel manifolds O(n)/O(n-2), spheres O(n)/O(n-1), and fixed points O(n)/O(n). The orbit space is a (m+3)-disk with the fixed point set S^m embedded in the boundary. With the natural smooth structure on the orbit space there is a singularity along the fixed point set.

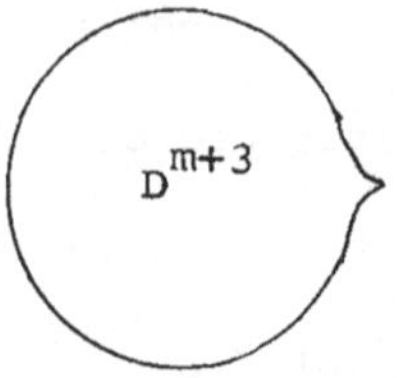

More generally, one can show that if Σ^{2n+m} is a biaxial $O(n)$-sphere, then the orbit space B has as similar structure,

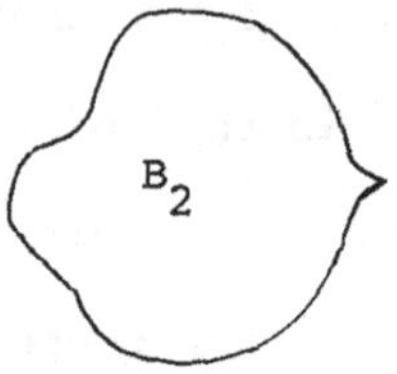

that is,

1) $B_2 = \operatorname{int} B$ is a contractible $(m+3)$-manifold, and

2) if n is even, $B_0 \underset{Z}{\sim} S^m$, (that is, B_0 is an integral homology sphere) while

2') if n is odd, the double branched cover of ∂B along B_0 is an integral homology sphere.

Conversely, the Hsiangs showed that given any such B, there is a unique biaxial $O(n)$-action on a homotopy sphere Σ^{2n+m} with $\Sigma/O(n) = B$ (as a stratified space).

So, for example, we could let $B = D^{m+3}$ and let B_0 be any knotted homotopy sphere in ∂B and obtain a unique $O(n)$-manifold with this as orbit space. Hence, the term "knot manifolds."

Some of these knot manifolds can also be realized as Brieskorn varieties (again, see [12]). For example, let

$$f(z) = (z_1)^p + (z_2)^q + (z_3)^2 + \ldots + (z_{n+2})^2$$

where p and q are odd and relatively prime, and let $\Sigma^{2n+1}(p,q) = f^{-1}(0) \cap S^{2n+3}$. Then $\Sigma^{2n+1}(p,q)$ is a homotopy sphere and supports a biaxial O(n)-action with orbit space D^4 and with the fixed point set a torus knot of type (p,q) in $\partial D^4 = S^3$. (See [12].)

Since the Brieskorn examples were known to have exotic differential structures, this raised the question of how the differential structure on Σ was related to the knot. It turned out that this relationship was the obvious one, that is, there are notions of Kervaire invariant and index for knots and if Σ corresponds to a knot, then it equivariantly bounds a parallelizable manifold of the corresponding index or Kervaire invariant. This result is due to Hirzebruch and Erle [9] and [7] for m = 1 and to Bredon [5] for m > 1. Also, roughly speaking,

$$\left\{\begin{array}{l}\text{concordance classes}\\ \text{of knot manifolds}\end{array}\right\} \longleftrightarrow \left\{\begin{array}{l}\text{knot cobordism}\\ \text{classes of knots}\end{array}\right\}.$$

This is true for n even, but for n odd we must replace knot cobordism by a "double branched cover knot cobordism" (analogous to condition 2'), above). More will be said about concordance in Chapter VI.

4. <u>U(n) and Sp(n) knot manifolds</u>: Suppose that d = 2 or 4 and that Σ^{2dn+m} is a biaxial $G^d(n)$-manifold. Let $B = \Sigma/G^d(n)$. Again it is not hard to show that

1) B is a contractible manifold of dimension m + d + 2,

and that

2) $B_0 \underset{Z}{\simeq} S^m$.

However, for $d \geq 2$, not every such space can occur as the orbit space of a biaxial $G^d(n)$-action. Also, at least when m = 1, a biaxial

$G^d(n)$-action ($d \geq 2$) on a homotopy sphere is not determined by its orbit space. (The negation of both of these facts was asserted in [14].) W. C. Hsiang and D. Erle in [11], found these mistakes and showed that such a B occurs as an orbit space if and only if B_0 bounds a parallelizable submanifold of ∂B. Bredon completed the classification of these U(n) and Sp(n) actions on homotopy spheres in 1973 in [6] by showing that the actions over a fixed orbit space are classified by framed cobordism classes of submanifolds of ∂B co-bounding B_0. (This result will be proved in III.D.)

References

[1] Borel, A. et al., Seminar on Transformation Groups, Ann. of Math. Studies 46, Princeton University Press (1961).

[2] Bredon, G., Fixed point sets and orbits of complementary dimension, Seminar on Transformation Groups, Ann. of Math. Studies 46, Princeton University Press (1961).

[3] __________, Examples of differentiable group actions, Topology 3, 115-122 (1965.

[4] __________, Exotic actions on spheres, "Proc. Conf. Transformation Groups", New Orleans 1967, 47-76. Springer-Verlag, Berlin and New York, 1968.

[5] __________, Introduction to Compact Transformation Groups, Academic Press, New York, 1972.

[6] __________, Regular O(n)-manifolds, suspension of knots, and knot periodicity, Bull. Amer. Math. Soc., 78(1973), 87-91.

[7] __________, Biaxial actions, preprint, (1973).

[8] Davis, M., Smooth G-manifolds as collections of fiber bundles. To appearl in Pac. Math. Journ. (1977).

[9] __________, W.C. Hsiang and W.Y. Hsiang, Differential actions of compact simple Lie groups on homotopy spheres and Euclidean spaces, to appear in the Proceedings of Stanford Topology Conf. (AMS).

[10] Erle, D., Die quadratische Form eines Knotens und ein Satz über Knotenmannegfaltigkeiten, J. Reine Agnew. Math. 236 (1969), 174-218.

[11] _______, and W.C. Hsiang, On certain unitary and symplectic actions with three orbit types, Amer. J. Math 94(1972), 289-308.

[12] Hirzebruch, F., Singularities and exotic spheres, Seminar Bourbaki 19(1966/67).

[13] Hsiang, W.C. and W.Y. Hsiang, Classification of differentiable actions on S^n, R^n and D^n with S^k as the principal orbit type, Ann. of Math. 82(1965), 421-433.

[14] _______, Differentiable actions of compact connected classical groups: I, Amer. J. Math. 89(1967), 705-786.

[15] _______, On compact subgroups of the diffeomorphism groups of Kervaire spheres, Ann. of Math. 85 (1967, 359-369.

[16] Hsiang, W.Y., On the classification of differentiable SO(n) actions on simple connected π-manifolds, Amer. J. Math. 88 (1966), 137-153.

[17] _______, A survey of regularity theorems in differentiable compact transformation groups. Proc. Conf. Transformation Groups, New Orleans 1967, 77-124, Springer-Verlag, Berlin and New York, 1968.

[18] Jänich, K., Differenzierbare Mannigfaltigkeiten mit Rand als Orbitraume differenzierbarer G-Mannigfaltigkeiten ohne Rand, Topology 5, 301-329.

[19] Montgomery D., and H. Samelson, Examples for differentiable group actions on spheres, Proc. N.A.S. 47, 1202-1205 (1961).

[20] Yang, C.T., Transformation groups on a homological manifold. Trans. Amer. Math. Soc. 87 (1958), 261-263.

II. SOME PRELIMINARY FACTS CONCERNING THE LINEAR MODEL AND ITS ORBIT SPACE

A. The linear representation $k\rho_n$.

Any information one can obtain concerning the orbit structure and orbit space of a linear representation will prove useful in the study of smooth transformation groups. Since we are interested in smooth actions with normal orbit types occurring among those of $k\rho_n$, it will help us to take a close look at this representation, and to see, at least, which normal orbit types we are considering.

First--some notation. Let $G(n)$ stand for either $O(n)$, $U(n)$ or $Sp(n)$ and let $\mathbb{F}$ stand for either $\mathbb{R}$, $\mathbb{C}$, or $\mathbb{H}$. Let $M(n,k)$ be the vector space of $(n \times k)$-matrices over $\mathbb{F}$, i.e., let

$$M(n,k) = \mathrm{Hom}_{\mathbb{F}}(\mathbb{F}^k, \mathbb{F}^n).$$

The natural action $G(n) \times M(n,k) \to M(n,k)$ (given by matrix multiplication) can be identified with $k\rho_n$ (i.e., with the direct sum of k copies of the standard representation). There is a natural $G(n)$-invariant inner product on $M(n,k)$, namely,

$$\langle x,y \rangle = \mathrm{tr}(x^*y)$$

where x^* is the conjugate transpose of x. In the next proposition we compute the normal orbit types of $G(n)$ on $M(n,k)$.

Proposition 1: Let $x \in M(n,k)$, let G_x be the isotropy group at x and let V_x be the normal representation at x. Then

1) G_x is the subgroup which fixes (pointwise) the subspace $\mathrm{Im}(x) \subset \mathbb{F}^n$.

2) $V_x = \mathrm{Hom}_{\mathbb{F}}(\ker(x), \mathrm{Im}(x)^{\perp})$. Here we are regarding $\mathrm{Hom}_{\mathbb{F}}(\ker(x), \mathrm{Im}(x)^{\perp})$ as a subspace of $M(n,k) \cong T_x(M(n,k))$.

Remark: If the rank of x is i, then $\mathrm{Im}(x) \cong \mathbb{F}^i$ and $\ker(x) \cong \mathbb{F}^{n-i}$ ($\mathrm{Im}(x)$ denotes the image of x). It follows that the pair (G_x, V_x) is equivalent to $(G(n-i), M(n-i,k-i))$.

The i-stratum of $M(n,k)$ (denoted by $M(n,k)_i$) is the submanifold of $M(n,k)$ consisting of those matrices x such that $\mathrm{rk}(x) = i$.

Proof of Proposition: The first statement is obvious. To prove the second, suppose that x has rank i. By the Slice Theorem, x has an invariant neighborhood of the form $G \times_{G_x} S_x$ where the slice representation S_x consists of those tangent vectors normal to the orbit. Let F_x be the subspace of S_x fixed by G_x, and recall that V_x is the orthogonal complement of F_x in S_x. Suppose that $y \in G \times_{G_x} F_x$. Then the G_y is conjugate to G_x; hence, the rank of y is also i. Conversely, if $y \in G \times_{G_x} (S_x - F_x)$ then $\mathrm{rk}(y) > i$. In other words, F_x is the subspace of S_x tangent to the i-stratum. It follows that V_x may be characterized as the subspace of the tangent space consisting of vectors normal to the stratum of x.

Now consider the exponential map $\exp\colon T_x(M(n,k)) \to M(n,k)$. This may be identified with the map $M(n,k) \to M(n,k)$ defined by $y \to x + y$. Suppose that $y \in \mathrm{Hom}(\mathbb{F}^k, \mathrm{Im}(x)) \subset M(n,k)$. Then for small values of y $\mathrm{Im}(x+y) = \mathrm{Im}(x)$, hence, y is tangent to the stratum $M(n,k)_i$. Next consider the case where $y \in \mathrm{Hom}(\ker x^{\perp}, \mathbb{F}^n)$. Then $y^* \in \mathrm{Hom}(\mathbb{F}^n, \mathrm{Im}(x^*))$. So, as above, for small values y, $\mathrm{Im}(x^* + y^*) = \mathrm{Im}(x^*)$. It follows that $\mathrm{rk}(y) = \mathrm{rk}(y^*) = \mathrm{rk}(x^*) = i$. Thus, y is again tangent to the stratum. Let E be the subspace of $M(n,k)$ spanned by $\mathrm{Hom}(\mathbb{F}^k, \mathrm{Im}(x))$ and $\mathrm{Hom}(\ker(x)^{\perp}, \mathbb{F}^n)$. We have shown that $V_x \subset E^{\perp} = \mathrm{Hom}(\ker(x), \mathrm{Im}(x)^{\perp})$. On the other hand, if $y \in \mathrm{Hom}(\ker(x), \mathrm{Im}(x)^{\perp})$, then $\mathrm{Im}(x + y) = \mathrm{Im}(x) + \mathrm{Im}(y)$. Hence, y is never tangent to the stratum. Therefore $V_x = \mathrm{Hom}(\ker(x), \mathrm{Im}(x)^{\perp})$. □

We now see that the isotropy groups of a k-axial $G(n)$-manifold are conjugate to $G(n-i)$ and the normal representations are equivalent to $(k-i)\rho_{n-i}$, where $i \leq \min(n,k)$.

<u>Definition</u>: Let M be a k-axial $G(n)$-manifold. The <u>i-stratum of M</u> (denoted by M_i) is defined as

$$M_i = \{x \in M \mid (G_x, V_x) \sim (G(n-i), M(n-i, k-i))\}.$$

Thus, the strata of M are linearly ordered and they can be indexed by the integers between 0 and $\min(n,k)$, inclusive. (We do not exclude that possibility that M_i is the empty set; however, notice that $M_i \neq \emptyset$ implies $M_j \neq \emptyset$ for $j > i$.) It follows from the Slice Theorem that M_i is a smooth invariant submanifold of M and that its image in the orbit space B is also a smooth manifold. Let us denote this image by B_i and call it the <u>i-stratum of B.</u> Also, notice that the above definition of the i-stratum of a $G(n)$-manifold coincides with our earlier definition of the i-stratum of $M(n,k)$.

B. <u>The orbit space of</u> $M(n,k)$.

Let $H(k)$ be the vector space of $(k \times k)$ $\mathbb{F}$-hermitian matrices (that is, as $\mathbb{F} = \mathbb{R}$, $\mathbb{C}$ or $\mathbb{H}$, $H(k)$ is either the set of symmetric, complex hermitian, or quaternionic hermitian matrices). Let

$H_+(k) \subset H(k)$ be the subset of positive semi-definite matrices. Consider the polynomial map $\pi: M(n,k) \to H(k)$ defined by

$$\pi(x) = x^*x.$$

Let us make a few observations:

1) If $n \geq k$, then the image of π is $H_+(k)$.

2) π is constant on orbits, i.e., $\pi(gx) = x^*g^{-1}gx = \pi(x)$. Hence there is an induced map $\bar{\pi}: M(n,k)/G(n) \to H_+(k)$.

3) If $n \geq k$, then $\bar{\pi}$ is a homeomorphism onto $H_+(k)$. (This essentially polar decomposition of matrices.)

4) In fact, $\bar{\pi}$ is a "smooth isomorphism" onto its image.

The proofs of 1) and 3) are left as exercises. To make sense of 4), we must consider what should be meant by a "smooth structure" on an orbit space. Just as the topology on M/G is defined by identifying the continuous functions on M/G with the G-invariant continuous functions on M, we should identify the "smooth" functions on M/G with the G-invariant smooth functions on M. In other words, $f: M/G \to \mathbb{R}$ is <u>smooth</u> if and only if $p \circ f$ is smooth, where $p: M \to M/G$ is the orbit map. This defined a "smooth structure" on M/G. (We do not mean, in any way, to imply that M/G is a smooth manifold or even a manifold.)

If V is a G-module, then we can also speak of the G-invariant polynomials on V. According to a theorem of G. Schwarz, the G-invariant smooth functions on V are related to the invariant polynomials in an obvious way. Before stating this theorem, let us recall a result of Hermann Weyl (page 52 in [7]).

Theorem 2 (Weyl): The entries of π: $M(n,k) \to H(k)$ generate the invariant polynomials on $M(n,k)$.

This theorem says that $H_+(k)$ is the orbit space of $M(n,k)$ in an algebraic sense. The next theorem, due to Schwarz [6], implies that it is also the orbit space in a C^∞ sense.

Theorem 3 (Schwarz): Suppose that V is a G-module (over $\mathbb{R}$) and that $\varphi_1,\ldots,\varphi_s$ generate the G-invariant polynomials on V. Let $\varphi = (\varphi_1,\ldots,\varphi_s)\colon V \to \mathbb{R}^s$. Then the induced map $\bar{\varphi}\colon V/G \to \mathbb{R}^s$ is a smooth isomorphism onto its image, i.e.

$$\varphi^* C^\infty(\mathbb{R}^s) = C^\infty(V)^G.$$

Notice that our map π: $M(n,k) \to H(k)$ is precisely the map φ in the hypothesis of Schwarz's theorem. In this special case, the result actually follows from an earlier theorem of G. Glaser [2].

The space $H_+(k)$ has very nice structure. Let us make a few observations about it.

1) If $x \in M(n,k)$ and $rk(x) = i$, then $rk(\pi(x)) = i$. Thus, the i-stratum of $H_+(k) = \{y \in H_+(k) \mid rk(y) = i\}$.

2) $H_+(k)$ is a convex cone.

3) $H_+(k)$ is homeomorphic (although not smoothly equivalent) to Euclidean half-space of dimension $k + 1/2\, dk(k-1)$, where $d = \dim_{\mathbb{R}} \mathbb{F}$.

Some Examples (of $H_+(k)$ for $k \leq 3$): $H_+(1)$ is the half-open ray $[0,\infty)$. $H_+(2)$ is the (open) cone on a disk of dimension $d+1$ (i.e., $H_+(2) \cong D^{d+1} \times [0,\infty)/(x,0) \sim *$.) $H_+(3)$ is a cone over D^{3d+2} where the

1-stratum is the cone over an $\mathbb{H}P^2$ embedded in ∂D^{3d+2} (there is a singularity along the $\mathbb{H}P^2$).

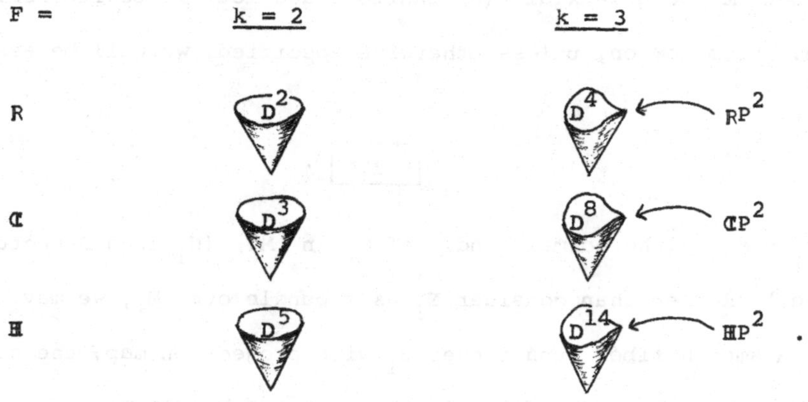

Facts about $H_+(k)$ immediately translate, via the Slice Theorem, into local information about orbit spaces of k-axial actions. For example, one sees from our description of $H_+(2)$, that the local structure of knot manifolds is as was asserted in the first lecture. As another example, we have the

<u>Proposition 4</u>: Let B be the orbit space of a k-axial $G(n)$-manifold, where $n \geq k$. Then B is <u>modeled on</u> $H_+(k)$ in the following sense.

a) B is a topological manifold with boundary.

b) If $x \in B_i$, then x has a neighborhood of the form $R^m \times H_+(k-i)$.

<u>Proof</u>: Suppose that $B = M/G(n)$. Pick $y \in M$ with $\pi(y) = x$. By the Slice Theorem, y has a neighborhood in M of the form $G \times_{G_y} S_y$, so x has a neighborhood in B isomorphic to $(G \times_{G_y} S_y)/G \cong S_y/G_y \cong F_y \times V_y/G_y \cong R^m \times H_+(k-i)$. This proves b). Statement a) follows from b) and observation 3) above. □

III. THE NORMAL BUNDLES OF THE STRATA AND TWIST INVARIANTS

A. The equivariant normal bundle of a stratum.

Let M be a k-axial $G(n)$ manifold and let B be its orbit space. From now on, unless otherwise specified, we will be assuming that

$$\boxed{n \geq k}\ .$$

Let $N_i \to M_i$ be the normal bundle of M_i in M. (N_i is a G-vector bundle.) Rather than consider N_i as a bundle over M_i, we may regard it as a smooth fiber bundle over B_i with projection map, the composition $N_i \to M_i \to B_i$. The fiber is the G-vector bundle $G \times_H V$, where $G = G(n)$, $H = G(n-i)$ and $V = M(n-i,k-i)$. The structure group S can be taken to be $\mathrm{Aut}_G(G \times_H V)$, the group of equivariant linear bundle automorphisms of $G \times_H V$. In general,

$$S = N_H(G \times GL(V))/H$$

where H is embedded diagonally in $G \times GL(V)$ (in $GL(V)$ via the normal representation). In our case,

$$S = G(i) \times GL(k-i),$$

where $GL(k-i) = GL(k-i;\mathbb{F})$. Here we regard $GL(k-i)$ as the centralizer of H in $GL(V)$ and $G(i)$ as the centralizer of H in G (also, $G(i) = N_H/H$). If $(g_1,g_2) \in G(i) \times GL(k-i)$ and $[g,v] \in G \times_H V$, then S acts from the right on $G \times_H V$ in the obvious fashion, namely, via $[g,v]\cdot(g_1,g_2) = [gg_1,vg_2]$. (For further details, see [1]. Also, there are similar results in [5].)

Now, let $P_i(M) \to B_i$ be the principal $G(i) \times GL(k-i)$ bundle associated to $N_i \to B_i$. $P_i(M)$ (= P_i) is called the i-normal orbit bundle of M. Notice that P_k is the principal G(k)-bundle associated to $M_k \to B_k$, that is, P_k is the principal bundle of the bundle of principal orbits. It follows from Lemma I.3 that $P_k(M) = F(G(n-k);M_k)$.

If we choose an invariant metric for N_i then the structure group of P_i can be reduced to $G(i) \times G(k-i) \subset G(i) \times GL(k-i)$. Given such a metric we, therefore, get a principal $G(i) \times G(k-i)$ bundle $E_i(M) \to B_i$ which can be regarded as a subset of $P_i(M)$.

B. The mysterious "twist invariants."

A smooth G-manifold can be decomposed as the union of equivariant tubular neighborhoods of the strata. In this way, the basic building blocks of a G-manifold are the normal bundles of the strata. In the case of multiaxial actions these normal bundles are determined by the P_i's (or the E_i's). There are many relationships between these normal orbit bundles. For example, we will soon show that if P_k is a trivial fiber bundle, then so is P_0, that is, the normal bundle of the fixed set is automatically trivial. (For k = 1, this was proved in Chapter I.) The most important of these relationships is the relationship between each E_i and P_k (= E_k), the bundle over the top stratum. This relationship can be expressed as the fact that E_i can be identified with a reduction of the structure group of E_k restricted to a certain subset. This reduction of the structure group is essentially what is meant by the "twist invariant" (in the Hsiangs' language) or the "characteristic reduction" (in the language of Jänich and Hirzebruch-Mayer). The twist invariant can be used to classify certain kinds of actions with two orbit types over a given orbit space (in particular, it classifies monoaxial actions and biaxial actions without fixed points). This is the classification theorem of the Hsiangs [3] and Jänich [4] referred to earlier. It can also be defined in a wider context (for example, for k-axial actions), and we shall do so below. However, in this general context, the relationship of twist invariants to the classification problem is not so obvious. The definitions given below will probably seem somewhat unreasonable. My only justification for them is that they work.

First consider the vector bundle N_i. Regarded as a $G(n)$-manifold it is k-axial. Hence, we can consider $P_k(N_i)$, the principal bundle of its bundle of principal orbits. We assert that

1) $P_k(N_i) \cong G(k) \times_{G(i)\times G(k-i)} P_i(M)$.

<u>Proof</u>: First notice that

$$P_k(M(n,k)) = F(G(n-k), M(n,k)_k)$$
$$= GL(k).$$

We regard $P_k(M(n,k))$ as a principal $G(k)$-bundle (i.e., as a left $G(k)$-space) and also as a right $GL(k)$-space (where $GL(k)$ = centralizer of $G(n)$ on $M(n,k)$). Next, consider $G \times_H V$, where as before $G = G(n)$, $H = G(n-i)$, $V = M(n-i,k-i)$. In a similar fashion, we have that

$$P_k(G \times_H V) = G(k) \times_{G(k-i)} GL(k-i).$$

Now,

$$N_i \cong (G \times_H V) \times_S P_i(M)$$

where $S = G(i) \times GL(k-i)$. Hence,

$$
\begin{aligned}
P_k(N_i) &= P_k(G \times_H V) \times_S P_i \\
&= G(k) \times_{G(k-i)} GL(k-i) \times_{G(i)\times GL(k-i)} P_i \\
&= G(k) \times_{G(i)\times G(k-i)} P_i .
\end{aligned}
$$

□

Next, pick an invariant metric for N_i. In other words, pick a reduction of the structure group of N_i to $G(i) \times G(k-i)$. In general, if $X \to A$ is a principal G-bundle, then equivalence classes of reductions of the structure group to K are in 1-1 correspondence with homotopy classes of sections of the bundle $X/K \to A$. Thus, our metric corresponds to a section $s: B_i \to P_i/G(i) \times G(k-i)$. According to equation (1), we have a $G(i) \times G(k-i)$ equivariant inclusion $P_i \subset P_k(N_i)$. Moreover,

$$P_i/G(i) \times G(k-i) = P_k(N_i)/G(k).$$

Thus, we have the following commutative diagram

$$
\begin{array}{ccc}
E_i & \hookrightarrow & P_i \subset P_k(N_i) \\
\downarrow & & \downarrow \quad \swarrow \\
B_i & \xrightarrow{s} & P_i/G(i) \times G(k-i).
\end{array}
$$

We introduce another principal $G(k)$-bundle $F_i = P_k(N_i)|_{s(B_i)}$. So, at this point we have a reduction of the structure group of F_i to $G(i) \times G(k-i)$ via $E_i \hookrightarrow F_i$.

If we choose an equivariant tubular map $\tau: N_1 \hookrightarrow M$, then we get an induced map of principal bundles $P_k(\tau): P_k(N_i) \hookrightarrow P_k(M)$. In this way, we can regard F_i as the restriction of $P_k(M)$ to $s(B_i)$.

Now to really see what's going on, let us make the following

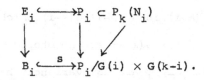

Hypothesis: $P_k(M)$ is a trivial fiber bundle.

Pick a trivialization $T: P_k \to G(k)$. Consider the following diagram

$$\begin{array}{ccccccc} E_i & \hookrightarrow & F_i & \hookrightarrow & P_k & \xrightarrow{T} & G(k) \\ \downarrow & & & & & & \downarrow \\ B_i & & & \longrightarrow & & & G(k)/G(i) \times G(k-i). \end{array}$$

Thus, we get a $G(i) \times G(k-i)$ equivariant map $\varphi_i: E_i \to G(k)$ covering $\bar{\varphi}_i: B_i \to G(k)/G(i) \times G(k-i)$. The homotopy class of $\bar{\varphi}_i$ is what is usually called the twist invariant.

Theorem 1: For each i, the equivariant homotopy class of $\varphi_i: E_i \to G(k)$ depends only on the homotopy class of the trivialization $T: P_k \to G(k)$.

Proof: The only choices involved in the definiton of φ_i are the metric, the tubular neighborhood and the trivialization T. Since the fiber of $P_i \to P_i/G(i) \times G(k-i)$ is $GL(k-i)/G(k-i)$ which is contractible, any two sections $B_i \hookrightarrow P_i/G(i) \times G(k-i)$ are homotopic. But this implies that the two maps $E_i \hookrightarrow P_i$ are equivariantly homotopic. It follows from the uniqueness part of the Equivariant Tubular Neighborhood Theorem, that the choice of tubular map $N_i \hookrightarrow M$ does not effect the equivariant homotopy class. Finally, it is clear that if we vary the trivialization $T: P_k \to G(k)$ by a $G(k)$-equivariant homotopy, we do not effect the homotopy class of φ_i.

□

Corollary 2: The homotopy class of $\bar{\varphi}_i: B_i \to G(k)/G(i) \times G(k-i)$ depends only on the homotopy class of the trivialization.

Thus, if M is a k-axial $G(n)$-manifold with trivial principal orbit bundle, we get invariants (of the equivariant diffeomorphism class of M).

$$[\bar{\varphi}_i] \in [B_i : G(k)/G(i) \times G(k-i)]/[B_k, G(k)].$$

A further corollary is the following.

<u>Corollary 3</u>: <u>Suppose</u> $G(n)$ <u>acts</u> k-<u>axially on</u> M <u>and that</u> $P_k(M)$ <u>is trivial. Then</u> E_i <u>is a pullback of the canonical bundle over a Grassmann (That is, of</u> $G(k) \to G(k)/G(i) \times G(k-i)$). <u>In particular</u>, E_0 <u>is the trivial bundle (it is the pullback of a bundle over a point)</u>.

C. <u>The classification of biaxial actions without fixed points</u>.

The twist invariant can be used to classify Bredon's examples of biaxial $O(n)$-actions with orbit space D^2. Suppose $O(n)$ acts biaxially on M^{2n-1} and $M/O(n) = D^2$. Then $P_2(M) = O(2) \times D^2$ and $E_1(M)$ is an $O(1) \times O(1)$ bundle over $\partial D^2 = S^1$. The twist invariant construction gives us

$$\begin{array}{ccc} E_1 & \xrightarrow{\varphi_1} & O(2) \\ \downarrow & & \downarrow \\ S^1 & \xrightarrow{\bar{\varphi}_1} & O(2)/O(1) \times O(1) \cong S^1. \end{array}$$

Thus, the homotopy class α of $\bar{\varphi}_1$ can be regarded as an element of $H^1(S^1;Z) = Z$. There are two homotopy classes of trivializations $O(2) \times D^2 \to O(2)$ ($O(2)$ has two components). It is not hard to see that the effect of changing the trivialization is to change the sign of α. Thus, $|\alpha|$ is a well-defined invariant of the action. In fact, we have the following,

<u>Proposition 4</u>: <u>Let</u> Σ_k^{2n-1} <u>be the biaxial</u> $O(n)$-<u>actions on the Brieskorn manifold defined by</u> $(z_1)^k + (z_2)^2 + \ldots + (z_{n+1})^2$. <u>Then the twist invariant of</u> Σ_k^{2n-1} is k.

We won't prove this, but it follows from the next theorem and a homology computation.

<u>Theorem 5</u> (Bredon, the Hsiangs, Janich): <u>There is a bijection</u>

$$\left\{\begin{matrix}\text{equivalence classes of } (2n-1) \\ \text{dimensional biaxial } O(n) \\ \text{manifolds with orbit space } D^2\end{matrix}\right\} \longleftrightarrow \{0, 1, 2, \ldots\} \quad .$$

<u>Proof</u>: First we show that we can construct an $O(n)$-manifold M, which realizes any twist invariant $\bar{\varphi}_1 : S^1 \to O(2)/O(1) \times O(1)$. Let E_1 be the pullback of the canonical $O(1) \times O(1)$ bundle over $O(2)/O(1) \times O(1)$ and let $\varphi_1 : E_1 \to O(2)$ be the map covering $\bar{\varphi}_1$. Set

$$M_2 = O(n)/O(n-2) \times D^2$$

$$N_1 = (O(n) \times_{O(n-1)} R^{n-1}) \times_{O(1)\times O(1)} E_1 .$$

Then φ_1 defines an equivariant diffeomorphism $f: \partial N_1 \to \partial M_2$. So, define $M = M_2 \cup_f N_1$. It is easy to check that twist invariant associated to M is $\bar{\varphi}_1$.

Secondly, we must show that if M and M' have the same twist invariant they are equivariantly diffeomorphic. We have an isomorphism $\lambda: P_2(M) \cong O(2) \times D^2 \cong P_2(M')$. Since $\bar{\varphi}_1$ and $\bar{\varphi}_2$ are homotopic, it follows from the Covering Homotopy Theorem that there is a bundle equivalence $\theta: E_1(M) \to E_1(M')$ such that the following diagram commutes up to homotopy

$$\begin{array}{ccc} E_1(M) & \xrightarrow{\varphi_1} & \partial P_2(M) \\ \theta \downarrow & & \downarrow \lambda \\ E_2(M') & \xrightarrow{\varphi_2} & \partial P_2(M') \end{array} .$$

Hence, we have an isomorphism $\times(\theta)\colon N_1(M) \to N_1(M')$ which agrees with the map induced by λ on ∂M_2. Putting $\times(\theta)$ and λ together we get an equivariant diffeomorphism $M \to M'$. □

The same type of result is true for biaxial $U(n)$ actions over D^3 and biaxial $Sp(n)$ actions over D^5. In the first case, we get a twist invariant $\partial D^3 \to U(2)/U(1) \times U(1) = S^2$ and in the second, we get $\partial D^5 \to Sp(2)/Sp(1) \times Sp(1) = S^4$.

D. Remarks concerning the general classification theorem.

I have a different use for twist invariants in mind. As usual, suppose that M is a k-axial $G(n)$-manifold and the $P_k(M)$ is trivial. One of the main theorems, which is proved (in outline) in the next chapter, is that there is a "stratified map" $f\colon M \to M(n,k)$. By a "stratified map" we mean that f is equivariant, strata-preserving, and that it maps the normal bundle of each stratum transversely. In fact, we will prove that such an f is unique up to a stratified homotopy and a choice of a homotopy class of trivialization of $P_k(M)$.

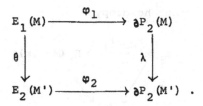

The first thing to observe is that since a stratified map f induces bundle maps $P_i(f)\colon P_i(M) \to P_i(M(n,k))$, the twist invariants are essentially functors in stratified maps (modulo the ambiguity introduced by the choice of trivialization of P_k). Thus, as a first obstruction to finding a stratified map we see that we must be able to find maps $f_i\colon E_i(M) \to E_i(M(n,k))$ so that the following diagram commutes up to equivariant homotopy

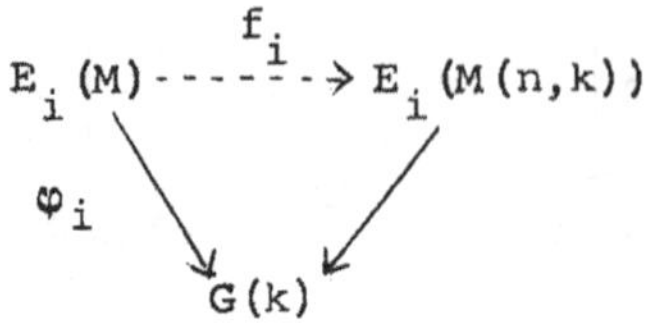

In the next lecture, I will show that $E_i(M(n,k)) = H_+(i)_i \times G(k)$ (where $G(i)$ acts by conjugation on $H_+(i)_i$ and by left multiplication on $G(k)$ and $G(k-i)$ also acts by left multiplication on $G(k)$). We will also show that the linear twist invariant $H_+(i)_i \times G(k) \to G(k)$ is projection on the second factor. Since $H_+(i)_i$ is contractible, this map is an equivariant homotopy equivalence. Thus, on each stratum there is, up to homotopy, essentially only one way to define the map f_i to make the above diagram commute. A slightly more careful analysis of this argument yields the result.

We will discuss a more precise version of this theorem and some of its applications in the next chapter.

References

[1] Davis, M., Smooth G-manifolds as collections of fiber bundles, (to appear in Pac. J. Math.).

[2] Glaser, G., Fonctions composees differentiables, Annals of Math. 77 (1963), 193-209.

[3] Hsiang, W.C. and W.Y. Hsiang, Differentiable actions of the compact connected classical groups: I, Amer. J. Math. 88(1966), 137-153.

[4] Jänich, K., Differenzierbare Mannigfaltigkeiten mit Rand als Orbiträume differenzierbarer G-Mannigfaltigkeiten ohne Rand, Topology 5, 301-329.

[5] __________, On the classification of O(n)-manifolds, Math. Ann. 176 (1968), 53-76.

[6] Schwarz, G., Smooth functions invariant under the action of a compact Lie group, Topology 14 (1975), 63-68.

[7] Weyl, H., The Classical Groups, second edition, Princeton University Press, Princeton, 1946.

IV. A STRUCTURE THEOREM FOR MULTIAXIAL ACTIONS AND SOME OF ITS CONSEQUENCES

A. The Structure Theorem.

1. The twist invariants of the linear model. Let M be a k-axial $G(n)$-manifold, with $n \geq k$. In the previous chapter we considered N_i, the normal bundle of the i-stratum, which could be regarded as bundle over B_i, the i-stratum of the orbit space. The structure group of this bundle can be reduced to $G(i) \times G(k-i)$. The associated principal bundle is denoted by $E_i(M) \to B_i$. The concept of a "twist invariant" was defined. Roughly speaking, a twist invariant arises from the fact that the principal orbits of N_i can be regarded as a bundle over the interior of $N_i/G(n)$ and also as a bundle over B_i. In other words, it arises from the fact that the G-manifold N_i has the finer structure of being a G-vector bundle.

To be a little more precise, the idea was to make use of an invariant inner product for N_i and a tubular neighborhood map, to push B_i into the interior of B. Since N_i is isomorphic to a tubular neighborhood of M_i, B_i has a neighborhood in B isomorphic to $N_i/G(n)$. That is to say, it has a neighborhood isomorphic to a fiber bundle with fiber $H_+(k-i)$. (Recall that $H_+(k)$ is the space of k by k positive semi-definite Hermitian matrices.)

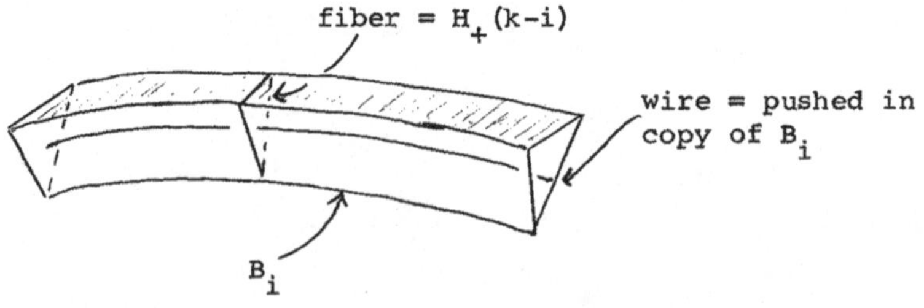

In the picture B_i is the bottom edge of the prism and the pushed in copy of B_i is the wire running through the center of the prism. We were able to identify E_i with a $G(i) \times G(k-i)$- equivariant subset of $E_k|_{wire} = F_i$. (Recall that E_k was the principal $G(k)$-bundle associated to the bundle of principal orbits $M_k \to B_k$.) If E_k is a trivial bundle, then so is $E_k|_{wire}$. Choosing a trivialization, $T: E_k \to G(k)$, we therefore had for each i, $0 \leq i \leq k$, a diagram,

$$\begin{array}{ccc} E_i & \xrightarrow{\varphi_i} & G(k) \\ \downarrow & & \downarrow \\ B_i & \xrightarrow{\bar{\varphi}_i} & G(k)/G(i) \times G(k-i). \end{array}$$

We showed that the equivariant homotopy class of $\varphi_i: E_i \to G(k)$ depends only on the equivariant homotopy class of the trivialization $T: E_k \to G(k)$. The homotopy class of $\bar{\varphi}_i$, which also depends only on the homotopy class of T, is a slight generalization of what the Hsiangs called the "twist invariant" (however, their definition is different from the above).

<u>Notation</u>: We shall use the following notation for the normal orbit bundles of the linear model:

$$E_i(k) = E_i(M(n,k)).$$

Also, we shall denote the base space of $E_i(k)$ by $B_i(k)$, that is to say,

$$\begin{aligned} B_i(k) &= \{\text{rank } i \text{ positive semidefinite hermitian matrices}\} \\ &= H_+(k)_i. \end{aligned}$$

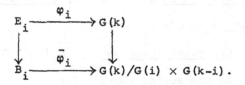

Proposition 1:

1) There is a canonical bundle equivalence

$$T_i: E_i(k) \longrightarrow B_i(i) \times G(k),$$

where $B_i(i) \times G(k)$ is regarded as a principal: $G(i) \times G(k-i)$ bundle as follows. Embed $G(i) \times G(k-i) \hookrightarrow G(k)$ is the standard fashion. $G(i)$ acts on $B_i(i) \times G(k)$ by conjugation on $B_i(i)$ (i.e., by $g \cdot x = g^{-1}xg$) and by left translation on $G(k)$. $G(k-i)$ acts trivially on $B_i(i)$ and by left translation on $G(k)$.

2) The linear twist invariant $\varphi_i: E_i(k) \to G(k)$ can be identified with projection on the second factor, i.e.,

$$\varphi_i = \text{proj} \circ T_i.$$

As a corollary to statement 1), we have the following interesting description of the base space of $E_i(k)$.

Corollary 2:

$$B_i(k) \cong B_i(i) \times_{G(i)} G(k)/G(k-i).$$

In other words, the space of $(k \times k)$ positive semidefinite hermitian matrices of rank i is diffeomorphic to a bundle with contractible fiber over the Grassmann of i-planes in k-space.

Proof of the Corollary: Although this follows immediately from statement 1) of the proposition, let us give a more intuitive proof. The corollary states that a $(k \times k)$-positive semidefinite hermitian matrix of rank i is determined by two pieces of data:

(i) an i-plane in $\mathbb{F}^k$,

(ii) an $(i \times i)$-positive definite matrix.

Let $x \in B_i(k)$. Then ker x is a $(k-i)$-plane and therefore $(\ker x)^{\perp}$ is an i-plane. Choosing a basis for $(\ker x)^{\perp}$, the restriction of x to $(\ker x)^{\perp}$ may be regarded as an $(i \times i)$ positive definite matrix. These two pieces of data clearly determine x. Moreover, the ambiguity introduced by the choice of basis, gives $B_i(k)$ the structure of the indicated bundle (rather than of a product). □

Example: It follows from this corollary that

$$B_1(k) = (0,\infty) \times \mathbb{R}P^{k-1}.$$

(Compare this with our description of $H_+(k)$ for $k \leq 3$ in Chapter II.B.)

Proof of the Proposition:

1) This is a matter of unraveling some definitions. Let N_i be the normal bundle of $M(n,k)_i$ in $M(n,k)$. We are interested in $E_i(k)$, which is the principal bundle associated to $N_i \to B_i(k)$. Let N_x be the fiber of $N_i \to B_i(k)$ at x. Then

$$N_x = G \times_{G_a} V_a,$$

where $G = G(n)$, $a \in \pi^{-1}(x)$ and V_a is the normal representation at a. According to Proposition II.1, $V_a \cong \mathrm{Hom}(\ker a, (\mathrm{Im}\, a)^{\perp})$. A point in $E_i(k)$ consists of a pair (x,θ) where $x \in B_i(k)$ and where $\theta\colon G(n) \times_{G(n-i)} M(n-i,k-i) \to N_x$ is an automorphism of Euclidean G-bundles (i.e., θ is an equivariant isometry). Any such θ can be written in the form $\theta([q,v]) = [qg^{-1}, L(v)]$, where $g \in G(n)$ is such that $g^{-1}G(n-i)g = G_a$ and where $L\colon M(n-i,k-i) \to V_a$ is an orthogonal isomorphism such that for all $k \in G(n-i)$ and for all $v \in M(n-i,k-i)$, we have that $L(kv) = (g^{-1}kg)L(v)$. (See Section 1.2 in [7] for further details.) Notice that the choice of $a \in \pi^{-1}(x)$ is completely arbitrary. There is a natural bundle equivalence $\lambda_g: G \times_{G_a} V_a \to G \times_{G_{ga}} V_{ga}$ defined by the pair $(g, dg: V_a \to V_{ga})$. Hence, if we replace a by $g^{-1}a$, θ is replaced by $\lambda_{g^{-1}} \circ \theta$. Thus, after making this substitution, we may assume that $G_a = G(n-i)$ and that θ is given by a pair $(1,L)$ where $L: M(n-i,k-i) \to V_a$ is an equivariant orthogonal isomorphism.

Now, we can define $T_i: E_i(k) \to B_i(i) \times G(k)$. First set

$$h = aa^* \in B_i(i).$$

Since h is positive definite, it has a unique positive definite square root. Let

$$u = h^{-\frac{1}{2}}a.$$

Then u is a linear transformation from $\mathbb{F}^n \to \mathbb{F}^i = \mathrm{Im}(a)$. Since $uu^* = h^{-1/2}hh^{-1/2} = 1$, we see that u is an orthogonal projection (the rows of u are an orthogonal basis for $(\ker u)^\perp = (\ker a)^\perp$).

Any equivariant orthogonal isomorphism $L: M(n-i,k-i) \to \mathrm{Hom}(\ker a, \mathbb{F}^{n-i})$ can be written in the form $L(v) = v \circ B^{-1}$ where $B: \ker a \to \mathbb{F}^{k-i}$ is an orthogonal isomorphism. Let $b = B \circ p$ where $p: \mathbb{F}^k \to \ker a$ is orthogonalprojection. Then $w = (u,b): (\ker a)^\perp \oplus \ker a = \mathbb{F}^k \to \mathbb{F}^k$ is an orthogonal isomorphism. Define $T_i: E_i(k) \to B_i(i) \times G(k)$ by

$$T_i(x,\theta) = (h,w).$$

It is easily checked that T_i is a $G(i) \times G(k-i)$ equivariant diffeomorphism. Notice that $G(k)$, regarded as the orthogonal centralizer of $G(n)$ on $M(n,k)$, acts on $E_i(k)$. It is also not difficult to see that T_i is equivariant with respect to this action on $E_i(k)$ and the natural (right) action of $G(k)$ on $B_i(i) \times G(k)$.

2) We wish to show that the twist invariant $\varphi_i: E_i(k) \to G(k)$ is just "projection on the second factor". The twist invariant is defined as the composition of several maps, the definitions of which depend on a choice of an invariant inner product for N_i and on a choice of tubular map. The standard inner product on $M(n,k)$ defines an inner product on N_i and using this metric we get an inclusion $I: E_i(k) \hookrightarrow E_k(N_i)$. If $\tau: N_i \hookrightarrow M(n,k)$ is a tubular map, then it induces a map on the k-stratum and consequently a map of the associated principal bundles $E_k(\tau): E_k(N_i) \to E_k(k)$. The twist invariant φ_i is a composition of maps as indicated in the following diagram,

$$\begin{array}{ccc} E_i(k) \overset{I}{\to} E_k(N_i) & \xrightarrow{E_k(\tau)} & E_k(k) \\ \varphi_i \downarrow & & \downarrow T_k \\ G(k) & \xrightarrow{\text{projection}} & B_i(i) \times G(k). \end{array}$$

First let us describe the map I. According to Lemma I.3 the principal bundle $E_k(N_i)$ can be regarded as a fixed point set, that is, $E_k(N_i) = F(G(n-k);(N_i)_k)$. Let $X = F(G(n-i);M(n,k)_i)$. It is not hard to see that the image of I lies in $F(G(n-k);(N_i)_k|_X)$ (actually, this last space can be identified with the bundle $P_i(M(n,k))$ defined in Chapter II). Let $(x,\theta) \in E_i(k)$ and let $a \in X$ and $B \in \mathrm{Hom}(\ker a,\mathbb{F}^{k-i})$ be as defined in the proof of 1). If V_a is the fiber of N_i at a, then $F(G(n-k),V_a) = \mathrm{Hom}(\ker a,\mathbb{F}^{k-i})$. Therefore, (a,B) may be regarded as a point in $F(G(n-k);(N_i)_k|_X) \subset E_k(N_i)$. A little checking through definitions shows that, indeed,

$$I(x,\theta) = (a,B).$$

Next, we need to define a tubular map $\tau\colon N_i \to M(n,k)$. We would like to use the exponential map, $\exp\colon N_i \to M(n,k)$; however, since this is only a diffeomorphism in some small neighborhood of the zero-section, we must scale it down. In other words, define τ on each fiber V_c by

$$\tau(y) = c + \epsilon_c(|y|)y$$

where $y \in V_c$ and where $\epsilon_c\colon [0,\infty) \to [0,\infty)$ is an appropriately chosen smooth function (i.e., ϵ_c = id near xero and ϵ_c has small values). Therefore

$$E_k(\tau)(a,B) = a + \epsilon b,$$

where $\epsilon = \epsilon_a(|B|)$. Now let us apply T_k to $A + \epsilon b \in E_k(k)$. First we calculate $H = (a + \epsilon b)(a + \epsilon b)^* = aa^* + \epsilon ab^* + \epsilon ba^* + \epsilon^2 bb^*$. Since $\ker a \perp \ker b$ and $\operatorname{Im} a \perp \operatorname{Im} b$, the cross terms vanish. Also, since b is an orthogonal projection $bb^* = 1 \in B_{k-i}(k-i)$. Thus, $H = aa^* + \epsilon^2$. Next we calculate

$$u = H^{-\frac{1}{2}}(a + \epsilon b)$$

$$= u + b = w$$

where u and w are as defined in the proof of 1). Then $T_k(a + \epsilon b) = (H,u)$. Putting all this together we have shown that the following diagram commutes,

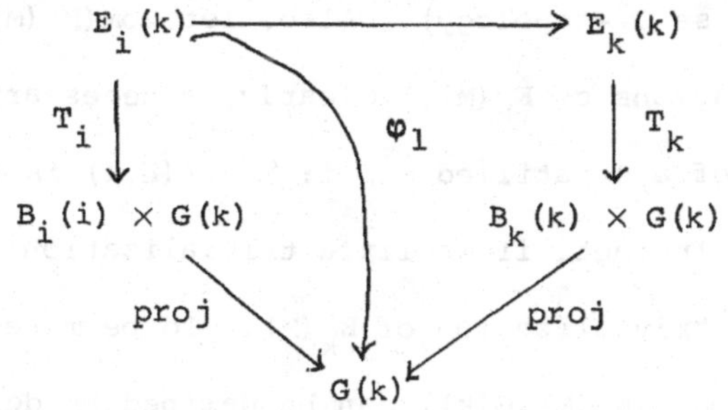

which is exactly what we wanted to show.

□

2. Stratified maps.

Definition: A smooth equivariant map $f\colon M \to M'$ of G-manifolds is stratified if

(i) $G_x = G_{f(x)}$;

(ii) the differential of f induces an isomorphism of normal representations $V_x \cong V_{f(x)}$.

Remarks: Stratified maps seem to be the "correct" morphisms for the category of smooth G-manifolds (see, for example, [7] and the paper of Browder and Quinn [5]). Notice that a stratified map induces a bundle map on each normal orbit bundle. Also notice that equivariant diffeomorphisms are stratified.

3. The Main Theorem: "You should start at the bottom (stratum) and work your way up." W. Browder.

The next theorem is one of our first examples of this important principle.

Suppose that M is a k-axial $G(n)$-manifold, with $n \geq k$. Let $\mathrm{Hom}(M,M(n,k))$ be the space of stratified maps from M to the linear model (with the coarse C^{∞}-topology). Also, let $\mathrm{Hom}(E_k(M),G(k))$ be the space of trivializations of $E_k(M)$. Clearly, a necessary condition for the existence of a stratified map $f: M \to M(n,k)$ is that the bundle $E_k(M)$ be trivial. In fact, if we fix a trivialization for $E_k(k)$, then f induces a trivialization of $E_k(M)$. To be more specific, $\Phi: \mathrm{Hom}(M,M(n,k)) \to \mathrm{Hom}(E_k(M),G(k))$ can be defined by defining $\Phi(f)$ by the diagram

$$\begin{array}{ccc} E_k(M) & \xrightarrow{E_k(f)} & E_k(k) \\ & {\scriptstyle \Phi(f)} \searrow \quad \swarrow {\scriptstyle \varphi_k} & \\ & G(k) & \end{array} .$$

Here, $E_k(f)$ is the bundle map induced by f and $\varphi_k = \mathrm{proj} \circ T_k$ is the linear twist invariant.

Theorem 3 (The Structure Theorem): *The map*

$$\Phi: \mathrm{Hom}(M,M(n,k)) \longrightarrow \mathrm{Hom}(E_k(M),G(k))$$

is a homotopy equivalence.

Remark: In particular, this theorem asserts that if $\mathrm{Hom}(E_k(M),G(k))$ is nonempty, then so is $\mathrm{Hom}(M,M(n,k))$. Thus, the obvious necessary condition for the existence of a stratified map is also sufficient. In practice, this crude version will be our only use of the theorem.

Sketch of the proof: We wish to construct a homotopy inverse for Φ, call it $\lambda: \mathrm{Hom}(E_k,G(k)) \to \mathrm{Hom}(M,M(n,k))$. That is to say, given a trivialization $T \in \mathrm{Hom}(E_k,G(k))$, we want to produce a stratified map

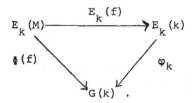

$f = \lambda(T): M \to M(n,k)$ in a fairly canonical way. The basic idea is simple. Given T, we can define twist invariants $\varphi_i: E_i \to G(k)$. Also, we have inclusions $s_i: G(k) \to B_i(i) \times G(k) = E_i(k)$ defined by $s_i(g) = (1,g)$, where $1 \in B_i(i)$ is the identity matrix. Let $\alpha_i = s_i \circ \varphi_i: E_i \to E_i(k)$. Then α_i defines a map of associated bundles $\times(\alpha_i): N_i \to N_i(k)$. The idea is that on a tubular neighborhood of the i-stratum the stratified map f will essentially be defined by $\times(\alpha_i)$. The only problem is that these maps agree only up to homotopy (and this is what makes the proof complicated). The actual proof proceeds, in outline, as follows.

<u>Step 1</u>: We first pick a "good" family of tubular neighborhoods and metrics for $M(n,k)$ and for M ("good" means that they match up correctly).

<u>Step 2</u>: Next, using these tubular maps, we adjust the trivialization T so that the following diagram commutes:

$$\begin{array}{ccc} E_k(N_i) & \xrightarrow{\text{proj}} & F_i = E_k|_{\text{wire}} \\ & {\scriptstyle T}\searrow \quad \swarrow{\scriptstyle T} & \\ & G(k) & \end{array} .$$

We can do this since proj: $E_k(N_i) \to E_k|_{\text{wire}}$ is a homotopy equivalence (the fiber is $B_{k-i}(k-i) \sim *$), and therefore this diagram always commutes up to homotopy. In fact, if we do this adjustment process carefully, it defines a homotopy equivalence $p: \text{Hom}(E_k, G(k)) \to A$ where A is the subspace consisting of those trivializations which satisfy the above diagram.

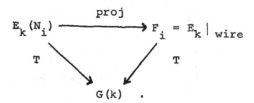

Step 3: Now we define the stratified map $f = \lambda(T): M \to M(n,k)$. The map f will be a bundle map on the prescribed tubular neighborhood of each stratum. In the construction we start from the bottom and work up (by induction). First we define f on a tubular neighborhood of the fixed set by $\times(\alpha_0): N_0 \to N_0(k)$. Next we try to extend this to the first stratum. Let $\bar{E}_1 = E_1 - E_1(N_0)$ and $\partial_0\bar{E}_1(\Sigma N_0)$ where ΣN_0 is the unit sphere bundle. Then α_0 defines a map $\times(\alpha_0): \partial_0\bar{E}_1 \to \partial_0\bar{E}_1(k)$, which we would like to extend to $\bar{E}_1 \to \bar{E}_1(k)$ by using α_1. The trouble is that $\alpha_1|_{\partial_0\bar{E}_1} \neq \times(\alpha_0)$; however, using Step 2, the following diagram commutes:

$$\begin{array}{ccc} \partial_0\bar{E}_1 & \xrightarrow{\times(\alpha_0)} & \partial_0\bar{E}_1(k) \\ \downarrow & & \downarrow \\ \bar{E}_1 & \xrightarrow{\varphi_1} & G(k) \end{array} .$$

It follows that

$$\begin{array}{ccc} \partial_0\bar{E}_1 & \xrightarrow{\times(\alpha_0)} & \partial_0\bar{E}_1(k) \\ \downarrow & & \downarrow \\ \bar{E}_1 & \xrightarrow{\alpha_1} & E_1(k) \end{array}$$

homotopy commutes. We can tack this homotopy onto a collared neighborhood of $\partial_0\bar{E}_1$ in $\bar{E}_1$ and then extend $\times(\alpha_0)$ to $\bar{E}_1$ via α_1. This new map $E_1 \to E_1(k)$ can then be used to define f on a tubular neighborhood of the 1-stratum. We continue in this fashion, making sure that during our homotopies we do not affect the commutativity of diagrams similar to the above, and in this way we define f.

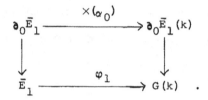

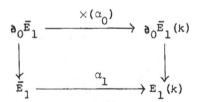

Step 4: Finally, we must show that Φ: Hom(M,M(n,k)) $\rightarrow$ Hom(E_k,G(k)) is a homotopy equivalence. We do this by showing that there are subspaces B $\subset$ Hom(M,M(n,k)) and A $\subset$ Hom(E_k,G(k)) such that the inclusions are homotopy equivalences and such that a $\Phi|_B$: B $\rightarrow$ A is a homeomorphism (with inverse $\lambda|_A$). The subspace A was described in Step 2. Roughly speaking, B consists of those stratified maps which are bundle maps on the prescribed tubular neighborhoods and which have the additional property that they collapse the complement of a collared neighborhood of ∂E_i to G(k) = the zero section of E_i(k). The verification that A and B have the desired properties is omitted.

Remark: In a certain sense (about which I do not want to be precise), the above theorem shows that the classifying space for k-axial G(n)-manifolds with trivial bundle of principal orbits is the linear model M(n,k).

The assumption that E_k(M) is a trivial fiber bundle is not really necessary. In fact, we could just as easily have proved the following

Theorem 4: Let Q be any smooth principal G(k) bundle and let X = M(n,k) $\times_{G(k)}$ Q. If M is any k-axial G(n)-manifold, with n $\geq$ k, then there is a homotopy equivalence Φ: Hom(M,X) $\rightarrow$ Hom(E_k(M),Q) defined as before.

In other words, if we can map E_k to Q, then we can map M to X. In parciular, if Q is the universal bundle EG(k) $\rightarrow$ BG(k), then the theorem asserts that any k-axial G(n)-manifold maps to

$$M(n,k) \times_{G(k)} EG(k).$$

Thus, this space can be regarded as the full classifying space for k-axial $G(n)$-manifolds with $n \geq k$.

B. The Pullback Construction and the Covering Homotopy Theorem:

In this section, I want to discuss two theorems, which will make life much easier for us, and which illustrate a close analogy between the theory of smooth G-manifolds and the theory of fiber bundles. Unlike our previous results, neither of these theorems is special to multiaxial actions.

By a local G-orbit space we shall mean a Hausdorff space B together with local charts to the orbit spaces of smooth G-vector bundles of the form $G \times_H S$ where S is an H-module, so that the overlap maps are smooth strata-preserving isomorphisms. It follows that a local G-orbit space has a "smooth" structure and a stratification by normal orbit types. Local G-orbit spaces are similar to what Thurston calls "orbifolds" and G. Schwarz called "Q-manifolds" in his thesis; however, we do not require any local lift of the overlap maps as part of the structure. We say that the local G-orbit space is modeled on $H_+(k)$ if the only vector bundles $G \times_H S$ which occur are of the form $G = G(n)$, $H = G(n-i)$, $S = M(n-i,k-i) \oplus R^m$. Notice that in this case, $(G \times_H S)/G \cong S/H \cong H_+(k-i) \times R^m$. Thus, B is modeled on $H_+(k)$ if it is locally isomorphic to $H_+(k-i) \times R^m$ (on each component of B, $\dim H_+(k-i) + m$ will be a constant).

Next we want to define a notion of "stratified map" of local G-orbit spaces which will mirror our definition of a stratified map of G-manifolds. First, we observe that once we have defined the smooth functions on B, we can define the tangent space $T_x(B)$ in the usual fashion. It follows from Schwarz's Theorem (which we stated in II.B) that $T_x(B)$ is a finite dimensional vector space of constant dimension along each stratum. Moreover, the "ordinary" tangent space to the stratum is a subspace. Hence, if B_α is a stratum of B, it is possible to define the normal bundle $W_\alpha \to B_\alpha$ the <u>normal bundle of</u> B_α <u>in</u> B. It is a vector bundle with fiber $T_x(B)/T_x(B_\alpha)$. A smooth strata-preserving map f: B → B' of local G-orbit spaces is <u>stratified</u> if for each stratum the induced map $W_\alpha(B) \to W_\alpha(B')$ is an isomorphism when restricted to each fiber. We should point out that an orbit space is obviously a local orbit space and that a stratified map of G-manifolds induces a stratified map of the orbit spaces.

<u>Remark</u>: These definitions are written down in much greater detail in [7]. However, there I use the terms "weak local G-orbit space" and "weakly stratified map" in place of the above terms.

The next theorem was first proved for biaxial actions by Bredon in [4]. In my thesis [6], I noticed that the same argument works for multiaxial actions; and in [7] I proved it for arbitrary smooth G-actions.

<u>Theorem 5</u>: <u>Suppose that</u> M <u>is a smooth</u> G-<u>manifold over</u> B <u>and that</u> B' <u>is a local</u> G-<u>orbit space. If</u> f: B' → B <u>is a stratified map, then the pullback</u>

$$f^*(M) = \{(x,y) \in M \times B' \mid \pi(x) = f(y)\}$$

<u>is a smooth</u> G-<u>manifold over</u> B'. <u>Moreover, the natural map</u> $f^*(M) \to M$

<u>is stratified and</u>

$$\begin{array}{ccc} f^*(M) & \longrightarrow & M \\ \downarrow & & \downarrow \\ B' & \longrightarrow & B \end{array}$$

<u>is a Cartesian square</u>.

<u>Remark</u>: We could have defined the subset $f^*(M) \subset M \times B'$ for any map f. It is a G-invariant subset of $M \times B'$ (where G acts by the given action on M and trivially on B'). The point of the above theorem is that if f is stratified, then $f^*(M)$ <u>will be a smooth manifold</u>. The proof uses Schwarz's result and the Implicit Function Theorem.

In the statement of the above theorem, the phrase "Cartesian square" means that $f^*(M)$ satisifes the usual universal property of pullbacks. That is to say, if M' is a G-manifold over B' and $F: M' \to M$ is a stratified map, then there exists a stratified map $\varphi: M' \to f^*(M)$ making the following diagram commute

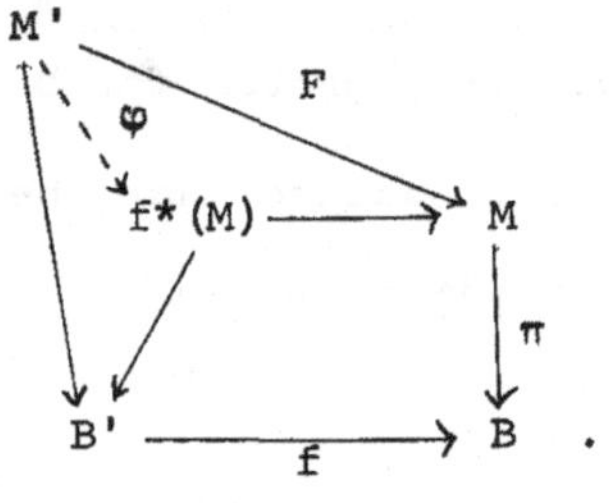

It follows from the fact that φ covers the identity on B', that φ is an equivariant diffeomorphism. (φ is a homeomorphism, since it is equivariant and maps each orbit to itself, and the condition that φ be stratified insures that φ is smooth and that the differential of φ is an isomorphism at each point.) Thus the relationship between stratified maps and pullbacks can be summed up in the following

Proposition 6: The statement that there exists a stratified map $M' \to M$ is equivalent to the statement that M' is equivariantly diffeomorphic to some pullback of M.

Next, I want to mention the following theorem of G. Schwarz, which is a smooth version of Palais, Covering Homotopy Theorem [15].

Theorem 7 (The Covering Homotopy Theorem): Let M and M' be smooth G-manifolds over B and B' and let $F: M' \to M$ be stratified. Suppose that $h: B' \times I \to B$ is a stratified homotopy with $h_0 = f$, the map induced by F. Then there is a stratified homotopy $H: M' \times I \to M$ extending F and covering h.

Remarks: Actually, using the pullback construction, this is a simple corollary of the "Covering Isotopy Theorem", which is the special case where $M' = M$, $F = id$, and h is an isotopy. This was conjectured by Bredon and proved by him in [3] for the case of monoaxial actions. I proved it for k-axial actions (with $n \geq k$) in my thesis [6]. Bierstone, in [1], proved it for actions of finite groups and also gave a different proof for biaxial $O(n)$-actions. By generalizing Bierstone's ideas on lifting vector fields, G. Schwarz [16] has recently proved the version stated above for arbitrary smooth actions of compact Lie groups.

Let us conclude this section with a little abstract nonsense. From now on, by a G-manifold we shall mean a triple (M,B,p) where G acts smoothly on M, B is a local G-orbit space, and $p: M \to B$ is a smooth map, constant on orbits, such that the induced map $\bar{p}: M/G \to B$ is an isomorphism. (This is analogous to the formalism of fiber bundle theory). Suppose that $\mathcal{b}$ is the category of G-manifolds and stratified maps and that $\mathfrak{B}$ is the category of local G-orbit spaces and stratified maps. There is a functor $\pi: \mathcal{b} \to \mathfrak{B}$ which associated B to M (B is the "base space"). The existence of pullbacks can be reformulated as a statement that "π is a fibered functor".

If B is any local orbit space, we can form the category $\pi^{-1}(B)$, called the fiber at B. The objects are G-manifolds with base space B and the morphisms are equivariant diffeomorphisms which induce the identity on B (such an equivariant diffeomorphism is called an <u>equivalence over B</u>). The notion of equivalence over B is completely analogous to the ordinary notion of equivalence of fiber bundles. The Covering Homotopy Theorem is essential to the study of $\pi^{-1}(B)$. It can be rephrased as follows: If f and g are homotopic maps from B to M/G, then $f^*(M)$ and $g^*(M)$ are equivalent over B.

C. A Classification Theorem.

Let B be a local G(n)-orbit space modeled on $H_+(k)$, and let

$$C_{n,k}(B) = \left\{\begin{matrix}\text{equivalence classes of } k\text{-axial} \\ G(n)\text{-manifolds over } B\end{matrix}\right\}.$$

Here, "equivalence classes" refers to equivalence over B. Next, let $\hat{C}_{n,k}(B)$ be the subset of $C_{n,k}(B)$ consisting of those M such that $E_k(M)$ is a trivial G(k)-bundle. Also, as usual, we suppose that $n \geq k$.

In the next theorem, which is a corollary of the Structure Theorem, we "calculate" $\hat{C}_{n,k}(B)$. We should think of this theorem as being analogous to the fact that fiber bundles are classified by homotopy classes, maps of the base space into the classifying space.

Theorem 8 (The Classification Theorem): There is a bijection

$$\hat{C}_{n,k}(B) \longleftrightarrow \pi_0 \mathrm{Hom}_{\mathfrak{B}}(B,H_+(k))/[B_k,G(k)].$$

Here, $\pi_0 \mathrm{Hom}_{\mathfrak{B}}(B,H_+(k))$ denotes the set of stratified homotopy classes of stratified maps $B \to H_+(k)$, while $[B_k,G(k)]$ is the group of ordinary homotopy classes of maps $B_k \to G(k)$.

Proof: By the Structure Theorem, every $M \in \hat{C}_{n,k}(B)$ is a pullback of $M(n,k)$ via some map $f\colon B \to H_+(k)$, the stratified homotopy class of which is unique up to the choice of homotopy class of trivialization of $E_k(M)$, that is, up to the action of $[B_k,G(k)]$ on $\pi_0 \mathrm{Hom}_{\mathfrak{B}}(B,H_+(k))$. Thus, $\hat{C}_{n,k}(B) \to \pi_0 \mathrm{Hom}_{\mathfrak{B}}(B,H_+(k))/[B,G(k)]$ is surjective. By the Covering Homotopy Theorem, it is also injective.

In the next lecture we will show that k-axial actions on spheres (or disks) have trivial k-normal orbit bundle and therefore the above theorem applies. Actually, the condition that $E_k(M)$ is trivial is only technical. In fact, using the second version of the Structure Theorem, one can prove the following

Theorem 9: Let $EG(k)$ be the universal $G(k)$-bundle and let

$$X = M(n,k) \times_{G(k)} EG(k).$$

Then there is a bijection

$$C_{n,k}(B) \longleftrightarrow \pi_0 \mathrm{Hom}_{\mathfrak{B}}(B,X/G)/G,$$

where G is the set of homotopy classes of bundle equivalences of principal $G(k)$-bundles over B_k.

We can use these theorems to answer two questions:

1) Given a local $G(n)$-orbit space B modeled on $H_+(k)$, is B actually the orbit space of some k-axial action (or of some k-axial action with trivial bundle of principal orbits)? By the above theorems, this is equivalent to the question of whether there exists a stratified map $B \to X/G$ (or $B \to H_+(k)$).

2) How many elements are there in $C_{n,k}(B)$ or in $\hat{C}_{n,k}(B)$?

D. Applications to Knot Manifolds and Other Biaxial Actions

First let us consider a biaxial $G(n)$ action on a homotopy sphere Σ of dimension $2dn + m$ (where $d = 1,2$ or 4 as $G(n) = O(n)$, $U(n)$, or $Sp(n)$). In the next chapter we shall show that the dimension of the fixed point set is m, and that each stratum of the orbit space B has the same homology (with appropriate coefficients) as the corresponding linear action on S^{2dn+m}. (In fact, we will show this is generally true for k-axial actions on spheres.) In particular, as was asserted in the first chapter,

1) B_2 is a contractible manifold of dimension $m + d + 2$;

2) $B_1 = \partial B - B_0 \underset{Z}{\simeq} S^d$;

3) $B_0 \underset{Z}{\simeq} S^m$.

In the $O(n)$ case, for n odd, statements 2) and 3) are only true with $Z/2$ coefficients; however, in this case the double branched cover of ∂B along B_0 will be an integral homology $(m+d+1)$-sphere.

First, we wish to consider an arbitrary local orbit space B

satisfying the above conditions and ask whether it is actually the orbit space of an action on a homotopy sphere. We can construct such a B by starting with a triple of smooth manifolds $(A, \partial A, A_0)$ where A is a contractible manifold with a homology sphere embedded in the boundary in codimension $d + 1$ and then by cutting out a tubular neighborhood of A_0 in A and pasting back in a bundle over A_0 with fiber $H_+(2)$ (so that there is a singularity along A_0).

Now suppose that we have a local orbit space B satisfying the above conditions. Since B is contractible, $\hat{C}_{n,2}(B) = C_{n,2}(B)$. The Classification Theorem asserts that B is an orbit space if and only if we can find a stratified map $B \to H_+(2)$. Moreover, $C_{n,2}(B)$ $\pi_0 \mathrm{Hom}(B, H_+(2))/[B_2, G(2)]$. Let us consider this in a little more detail.

<u>The O(n)-case</u>: In this case $H_+(2)$ is a cone on a 2-disk. What does it mean to find a stratified map $B \to H_+(2)$? First of all, we must map B_0 to $0 \in H_+(2)$. Next we must map a tubular neighborhood of B_0 into $H_+(2)$. This is essentially a trivialization of the normal bundle of B_0. Since $B_0 \subset \partial B$ is a knot of codimension two, it is always possible to find such a trivialization (B_0 has a neighborhood isomorphic to $B_0 \times H_+(2)$). Next remove the 1-stratum of this tubular neighborhood from B_1 and let $\partial_0 B_1$ be the boundary of the complement. Then $\partial_0 B_1$ is a circle bundle over B_0. Our trivialization defines a map $\partial_0 B_1 \to \partial_0 B_1(2) \cong S^1$. To construct a stratified map we must be able to extend this map to $B_1 \to B_1(2) \cong S^1 \times [0,\infty)$. If we can find such an extension, then we are done, for there is no problem in extending first to a collared neighborhood of B_1 and then over all of B_2 (since

the target in this last extension is $B_2(2)$ which is contractible). This whole procedure should be reminiscent of one construction of a Seifert surface for a knot. In fact, the problem of extending the map on $\partial_0 B_1$ (to B_1) is equivalent, via the Pontriagin-Thom construction, to the existence of a Seifert surface for B_0. Since such Seifert surfaces always exist for knots in codimension two, we can always find the extension to B_1 (possibly after altering the trivialization of the normal bundle of B_0). Therefore, given a local orbit space B, the strata of which have the correct homology, we can always find a stratified map $f: B \to H_+(2)$. It is then not difficult to prove that $f^*(M(n,k))$ is a homotopy sphere.

We should also notice that stratified homotopy classes of maps $B \to H_+(2)$ correspond, via the Pontriagin-Thom construction, to framed cobordism classes of Seifert surfaces cobounding B_0. In fact, any two Seifert surfaces are framed cobordant (this can be seen, for example, by a relatively simple homotopy theoretic argunent). Therefore, there is a unique biaxial action over any such B. (This is the theorem of the Hsiangs and Jänich which we discussed in the first lecture.)

The U(n) and Sp(n) cases: In these cases $H_+(2)$ is either a cone over D^3 or a cone over D^5. Thus, we are essentially dealing with knots of codimension three of five. The analysis of the problem of existence of a stratified map is similar to the O(n) case; however, the answer is different. First of all, in the case of codimension five knots, I don't believe that the normal bundle is necessarily trivial. Even if it is, we cannot always find an extension to

$B_1 \to B_1(2) = S^d \times (0,\infty)$. (In other words, knots of codimension greater than two do not necessarily have Seifert surfaces!) We might mention that this is almost exactly how Levine analyzed knots in codimension greater than two in [12]. In particular, for knots of codimension k he derived an exact sequence,

$$P_{n+1} \longrightarrow \theta^{m,n} \longrightarrow \pi_n(G_k, SO_k) \longrightarrow P_n$$

where P_{n+1} is the simply connected surgery group $\theta^{m,n}$ is the group of knotted n-spheres in S^m, $k = m - n$, and the map $\theta^{m,n} \to \pi_n(G_k, SO(k))$ is precisely the obstruction to finding a stratified map. Although this sequence was originally defined for isotopy classes of knotted homotopy n-spheres, it is also valid if we interpret $\theta^{m,n}$ as homology h-cobordism classes of knotted homology n-spheres.

For further details on the above, see Bredon's paper [4]. The above discussion constitutes an outline of the following result of Bredon (although his proof is slightly different).

<u>Theorem 10</u> (Bredon): <u>Let</u> B <u>be a local</u> $G(n)$-<u>orbit space modeled on</u> $H_+(2)$. <u>Then there is a bijection</u> $\hat{C}_{n,2}(B) \longleftrightarrow \mathcal{S}/[B_2, G(2)]$, <u>where</u>

$$\mathcal{S} = \left\{\begin{array}{l}\text{framed cobordism classes of submanifolds} \\ \text{of } \partial B \text{ cobounding } B_0\end{array}\right\}.$$

Notice that if B_2 is contractible, then $[B_2, G(2)]$ is the trivial in the U(n) and Sp(n) cases and is isomorphic to Z/2 in the O(n) case.

<u>Remark</u>: The original proofs of the Hsiangs [9] and Jänich [10] in the O(n)-knot manifold case was different from the above. Their idea

was to use the classification theorem for actions with two orbit types to classify the part of the knot manifold lying over $B - B_0$. As we showed in the second lecture, such actions are classified by their twist invariant in $H^1(B_1,Z)/Z_2 \cong \{0,1,2...\}$. Since the action must be equivalent to the product action on $D^{2dn} \times B_0$ = tubular nbhd of B_0, the twist invariant must be 1. They then studied the problem of gluing the tubular neighborhood of the fixed point set back in. It turned out, rather fortuitously, that for O(n) there was essentially only one way to do this; hence, the theorem follows. In the U(n) and Sp(n) cases the analysis of this gluing (i.e., of equivariant diffeomorphisms of the normal sphere bundle) proved to be more difficult (see [4], [6]).

This type of approach is an example of "starting from the top and working down". It is analogous to trying to do obstruction theory by defining the map on the cells of highest dimension and then trying to extend across the lower skeleta. It should be clear that as the number of strata increases, this approach leads to insurmountable difficulties. On the other hand, the proof I outlined above using the Classification Theorem is an example of "starting from the bottom and working up".

E. Applications to Triaxial Actions without Fixed Points.

Here, we are interested in G(n)-actions modeled on $3\rho_n$ on homotopy spheres of dimension 3dn - 1. The orbit space B has 3 non-empty strata. (It follows from ths assumption on the dimension that the fixed point set is empty.) In the next lecture we shall show that

1) B_3 is a contractible manifold of dimension 3d + 2;

2) $B_2 = \partial B - B_1 \underset{Z}{\sim} \mathbb{H}P^2$;

d) $B_1 \underset{Z}{\sim} \mathbb{H}P^2$.

(Recall that the symbol $X \underset{Z}{\sim} Y$ means that X and Y have the same integral homology.) In the O(n) case, for n even, there is the additional requirement that the double branched cover of ∂B along B_1 is an integral homology $\mathbb{C}P^2$.

We first investigate the problem of finding a stratified map $B \to H_+(3)$. Since B has no 0-stratum, we may assume that the image of B is contained in the image of the unit sphere in $H_+(3)$, that is, that the image is in

$$A = \{x \in H_+(3) \mid tr(x) = 1\}.$$

If B is any local orbit space modeled on $H_+(3)$, then it can be shown that the 1-stratum has a neighborhood in $B - B_0$ isomorphic to a fiber bundle over B_1 with fiber $H_+(2)$ and with structure group O(2) (acting by conjugation on $H_+(2)$). This bundle is denoted by $C_1(B)$ and is called the <u>normal cone bundle of</u> B_1. The first question is whether $C_1(B)$ is a pullback of $C_1(A)$, the normal cone bundle in the linear model. Just as in the knot manifold case, this reduces to a question about the normal bundle of B_1 in ∂B. This question was investigated and solved by Massey in [14].

<u>Proposition 11</u> (Massey): <u>Suppose</u> $X^{2d} \underset{Z}{\sim} \mathbb{H}P^2$. <u>If</u> d = 4, <u>also suppose that the first Pontriagin class</u> $p_1(X) = 2\cdot$<u>generator of</u> $H^4(X)$. <u>Let</u> $X \hookrightarrow Y^{3d+1}$ <u>be any embedding of</u> X <u>in an integral homology sphere</u> Y. <u>Then the normal bundle of</u> X <u>in</u> Y <u>is independent of the embedding</u>

and of the homology sphere.

Remark: The normal bundle of X in Y is stably equivalent to the stable normal bundle of X. Massey showed that it was determined (unstably) by certain characteristic classes described below.

1) If d = 1, the normal bundle of $\mathbb{R}P^2 \hookrightarrow Y^4$ is determined by the fact that its first Stiefel-Whitney class is nonzero and by the fact that its twisted Euler class is $2 \cdot$generator of $H^2(\mathbb{R}P^2, Z^{\text{twisted}})$.

2) If d = 2, the normal bundle of a homology $\mathbb{C}P^2$ in a homology 7-sphere is determined by the fact that $w_1 = 0$, $w_2 \neq 0$ and $p_1 = -3u^2$, where u is the generator of $H^2(X)$.

3) If d = 4, the bundle is determined by the Pontriagin classes of X. Therefore, in all three cases, given B, we can find a bundle map $C_1(B) \to C_1(A)$ covering a degree-one map $B_1 \to A_1 = \mathbb{F}P^2$.

Remark: The Pontriagin classes of the quaternionic projective plane are given by $p_1 = 2u$ and $p_2 = 7u^2$, where u is the generator of $H^4(\mathbb{H}P^2, Z)$. An arbitrary integral homology $\mathbb{H}P^2$,X may have quite different Pontriagin classes (although they are still subject to a relation from the Index Theorem, namely, $(7p_2 - p_1^2)[X] = 45$. On the other hand, it is not hard to see that if $C_1(B)$ pulls back from $C_1(A)$, then $p_1(B_1) = 2 \cdot$generator of $H^4(B_1)$. Therefore, a necessary condition for a homology $\mathbb{H}P^2$,X to occur as the 1-stratum of the orbit space of a triaxial action on a homotopy sphere is that $p_1(X) = 2 \cdot$generator. Thus, not every such X can occur. One can also show that this is a sufficient condition.

A bundle map $C_1(B) \to C_1(A)$ restricts to a map $\partial_1 B_2 \to \partial_1 A_2$. In order to findaa stratified map $B \to A$, we must extend this to a map $B_2 \to A_2$ (by Corollary 2 in section A, A_2 is homotopy equivalent to $\mathbb{H}P^2$). This extension problem turns out to be very difficult. If we could solve it, however, there is no problem in finishing the construction of the stratified map (since A_3 is contractible). In the $O(n)$ and $U(n)$ cases one can show that such an extension always exists. The answer in the $Sp(n)$ case is unknown. In fact, Massey wrote to me that he didn't believe anyone had ever solved a homotopy theoretic problem of this level of difficulty. Thus, for $O(n)$ and $U(n)$ there is the following theorem (which I proved in my thesis).

<u>Theorem 12</u>: <u>Suppose</u> $G(n) = O(n)$ <u>or</u> $U(n)$. <u>Let</u> B <u>be a local orbit space satisfying the conditions on the homology of its strata given at the beginning of this section. Then there exists a stratified map</u> $f\colon B \to A \subset H_+(3)$. <u>Moreover, the pullback is a homotopy sphere</u>.

Finally, we come to the problem of calculaing $\hat{C}_{n,3}(B)$, i.e., of computing the homotopy classes of stratified maps $B \to A$. Again, this turns out to be tractable only in the $O(n)$ and $U(n)$ cases. For $O(n)$, $\hat{C}_{n,3}(B) \cong Z/2 \times H$, where H is the torsion in the second cohomology of the double branched cover of ∂B. For $U(n)$, I am about 90 percent certain that $\hat{C}_{n,3}(B)$ is a singleton. Let us finish this section with a

<u>Problem</u>: In the case of triaxial actions without fixed points find some interpretation of $\pi_0 \mathrm{Hom}(B,A)$ analogous to its interpretation as framed cobordism classes of Seifert surfaces in the knot manifold case.

F. Some "Counterexamples."

Of course it is not generally true that every smooth G-action on a homotopy sphere which is modeled on some linear representation is the pullback of its linear model. For example, it is not true for k-axial G(n) actions when $k > n$. (Our proof certainly breaks down since the twist invariants are not even defined.) The simplest example of this failure is the following.

Biaxial U(1)-actions. Here we are considering semifree S^1-actions with fixed point set of codimension four. Levine [13] studied such actions on homotopy spheres Σ^{m+4}. He showed that the orbit space B was a homotopy $m + 3$ sphere with fixed point set B_0 a homology m-sphere. Moreover, he showed that any such pair (B,B_0) is the orbit space of a unique action on a homotopy sphere. But we cannot always find a stratified map from B to the linear model. For, as we showed in Section D, this would imply that B_0 has a Seifert surface in B and there are knots in codimension three without this property. Thus, there are biaxial U(1)-actions on spheres which do not pull back from $M(1,2) = \mathbb{C}^2$. In particular, this implies that there are biaxial U(1) actions on spheres which do not extend to biaxial U(2)-actions.

The question arises: Does the Structure Theorem hold for k-axial SO(n)-actions and k-axial SU(n)-actions? It turns out that for $n \geq k + 2$ the answer is yes. In fact, as Jänich observed in [11], in this range there is no difference between k-axial SO(n)-actions (respectively, SU(n)) and k-axial O(n)-actions (respectively, U(n)), that is to say, for $n \geq k + 2$ every k-axial SO(n)-action extends to a unique k-axial O(n)-action. From another angle, the proof of the Structure Theorem is valid for SO(n), since as before we can define the normal orbit bundles E_i with structure groups $O(i) \times O(k-i)$.

However, for k = n or n - 1 the situation is completely different. This is certainly not surprising when k = n; for then in the linear model $SO(n) \times M(n,n) \to M(n,n)$ we can really not distinguish between the n-stratum and the (n-1)-stratum(i.e., $G_x = \{1\}$ if $rk(x) = n$ or $rk(x) = n - 1$). For k = n - 1 the reason is much more subtle. In this case the bundle E_k has structure group SO(n), while in the O(n) case it has structure group $O(k) = O(n-1)$ (this discrepancy essentially arises from the fact that $SO(1) = \{1\}$ but $O(1) \neq \{1\}$). Just as before we can go ahead and assume that the bundle of principal orbits is trivial and define the twist invariants. However, the i-twist invariant will lie in some quotient of SO(n), while the i-stratum of the linear orbit space will still be homotopy equivalent to the Grassmann $O(n-1)/O(i) \times O(n-1-i)$. Thus, there is no reason to expect (n-1)-axial SO(n)-actions to all pull back from the linear model (and in fact they don't). Similar remarks apply to SU(n).

Biaxial SO(3)-actions. This discrepancy is already apparent in the study of biaxial SO(3)-actions (as Bredon has observed). First of all, in his 1965 paper [2], Bredon showed that any two biaxial SO(3)-actions on S^5 were equivalent. To be more precise, there are only two equivalence classes of SO(3)-actions on the Brieskorn manifolds Σ_k^5 defined by $(z_1)^k + (z_2)^2 + (z_3)^2 + (z_4)^2$ depending on whether k is even or odd (however, as O(3)-actions they are inequivalent for distinct nonnegative values of k). Since Σ_3^5 is the Kervaire sphere, this gives a completely transformation group theoretic proof that the 5-dimensional Kervarie sphere is standard.

The case of biaxial SO(3)-actions with fixed points is even more interesting. Bredon told me about the following example of Gromoll and Meyer [8]. Consider Sp(2) as a subset of 2 × 2 quaternionic matrices. Let Sp(1) act on Sp(2) by left multiplication on the first column and conjugation on the second column. They show that the orbit space Σ^7 is the generator of the group of exotic 7-spheres. They also observe that the centralizer of this action in Sp(2) is SO(3) × O(2). Hence SO(3) × O(2) acts smooth on Σ. The restriction of this action to SO(3) is biaxial. Moreover, Σ/SO(3) is a 4-disk with Σ_0 an unknotted circle in the boundary. It follows that Σ _cannot be a pullback of the linear_ SO(3)-_action on_ M(3,2). For if it were, then the action would extend to O(3) (with Σ/SO(3) = Σ/O(3)). But by the classification of knot manifolds any O(3) action with this orbit space is equivalent to the linear action on S^7, which is impossible since $\Sigma^7 \neq S^7$. However, there is a theorem for biaxial SO(3)-actions which is analogous to the Structure Theorem. Consider the quaternionic projective plane $\mathbb{H}P^2$. Since the group of automorphisms of $\mathbb{H}$ is SO(3), the groups act on $\mathbb{H}P^2$. The action is biaxial and the fixed point set is $\mathbb{R}P^2$. The orbit space X is homeomorphic to a 5-disk. As a stratified space $X \cong W_+(3) = \{x \in H_+(3) \mid tr(x) = 1\}$. This is also the orbit space of S^{11} by the linear triaxial SO(4) action; however, the strata have a different indexing set. Moreover, the following theorem is true.

Theorem 13: _Let_ B _be a local orbit space modeled on_ $H_+(2)$. _Then_ every biaxial SO(3)-action over B with trivial bundle of principal orbits is a pullback of $\mathbb{H}P^2$. In fact, such SO(3) manifolds are bijective with $\pi_0 \mathrm{Hom}(B,X)/[B_2, SO(3)]$.

Remark: In light of this theorem the fact that k-axial O(n)-actions pull back from M(n,k) seems more coincidental.

Question: What is the classifying space for (n-1)-axial SO(n)-actions (n > 3) with trivial bundle of principal orbits?

For further examples of biaxial SO(3)-actions on homotopy 7-spheres and for a further discussion of the above theorem, see Appendix 1.

G. Further Classification Theorems:

Suppose that $G = \Pi G_i$ where each G_i is of the form $O(n_i)$, $U(n_i)$ or $Sp(n_i)$. Further, suppose that $\varphi = \Pi \varphi_i$ is a representation, where each $\varphi_i = k_i \rho_{n_i}$ with $n_i \geq k_i$. Then analogous version of the Structure Theorem and the Classification Theorem are true for smooth G-manifolds modeled on φ. The proof is essentially the same.

For example, consider T^n actions modeled on the standard representation of T^n on $\mathbb{C}^n$ $(T^n = U(1)^n)$. Then all such T^n actions with trivial bundle of principal orbits pull back from the linear action on $\mathbb{C}^n$.

The method of twist invariants is also applicable to actions of disjoint and near adjoint type. These are actions modeled on certain irreducible representations (such as the adjoint representation). Twist invariants can be defined for these actions, and many times can be used to prove that these actions pull back from their linear models. These theorems will be discussed in a future joint paper with the Hsiangs.

References

[1] E. Bierstone, Lifting isotopies from orbit spaces, Topology, (1975), 245-252.

[2] G. Bredon, Transformation groups on spheres with two types of orbits, Topology, 3 (1965), 103-113.

[3] ________, Introduction to Compact Transformation Groups, Academic Press, New York, 1972.

[4] ________, Biaxial actions, mimeographed notes, Rutgers University, 1973.

[5] W. Browder and F. Quinn, A surgery theory for G-manifolds and stratified sets, Manifolds, University of Tokyo Press, 1973, 27-36.

[6] M. Davis, Smooth actions of the classical groups, Thesis, Princeton University, 1974.

[7] ________, Smooth G-manifolds as collections of fiber bundles, to appear in Pac. J. of Math.

[8] D. Gromoll and W. Meyer, An exotic sphere with nonnegative sectional curvature, Ann. of Math., 100 (1974), 447-490.

[9] W.C. Hsiang and W.Y. Hsiang, Differentiable actions of compact connected classical groups: I. Amer. J. Math., 89 (1967), 705-786.

[10] K. Jänich, Differenzierbare Mannigfaltigkeiten mit Rand als Orbitraume differenzierbarer G-Mannigfaltigkeiten ohne Rand, Topology, 5 (1966), 301-329.

[11] ________, On the classification of O(n)-manifolds, Math. Ann., 176 (1968), 53-76.

[12] J. Levine, A classification of differentiable knots, Ann. of Math., 82 (1965), 15-50.

[13] ________, Semifree circle actions on spheres, Inv. Math., 22 (1973), 161-186.

[14] W.S. Massey, Imbeddings of projective planes and related manifolds in spheres, Indiana Math. J., 23 (1974), 791-812.

[15] R.S. Palais, The Classification of G-spaces, Mem. Amer. Math. Soc., 36 (1960).

[16] G. Schwarz, Covering smooth homotopies of orbit spaces, to appear.

V. ACTIONS ON SPHERES

In the previous chapter we sketched a proof of a Structure Theorem stating that if M is a k-axial $G(n)$-manifold ($n \geq k$) with trivial bundle of principal orbits, then M is a pullback of the linear model $M(n,k)$. It was further asserted that any k-axial action on a homotopy sphere Σ has trivial bundle of principal orbits $\Sigma_k \to B_k$; and hence, that the Structure Theorem applies. This assertion will be proved in this chapter showing that B_k is, in fact, contractible. Along the way we will verify the orbit space of Σ is in a certain sense "homologically modeled" on the orbit space of the linear k-axial action on the sphere of the same dimension (as we asserted in the previous chapter in sections IV.D and IV.E). More precisely, at least for $G(n) = U(n)$ or $Sp(n)$, one can give necessary and sufficient conditions on a stratified map $f: B \to H_+(k)$ so that the pullback will be a homotopy sphere (or a disk). For example, for $G(n) = U(n)$ or $Sp(n)$ the pullback will be a homotopy sphere if and only if the following conditions are satisfied.

1) B is simply connected,

2) B_0 is an integral homology sphere, and

3) For $i > 0$, $f|_{B_i}: B_i \to H_+(k)_i$ induces an isomorphism on integral homology.

These results should be thought of as being analogous to the facts that any free circle action on M^{2n+1} is the pullback of the linear action on S^{2n+1} via a map $g: M/S^1 \to \mathbb{C}P^n$ and that the pullback is a homotopy sphere if and only if g is a homotopy equivalence.

A. The homology of the strata

1. The $U(n)$ and $Sp(n)$ cases. The result which we will prove in this section is not only more difficult to prove for k-axial $O(n)$-manifolds but also more difficult to state. For the unitary and symplectic groups the result is the following.

Main Lemma 1: Let $G(n) = U(n)$ or $Sp(n)$. Let $F: M \to M'$ be a stratified map of k-axial $G(n)$-manifolds covering $f: B \to B'$. Then F induces an isomorphism on integral homology if and only if $f|_{B_i}$ induces an isomorphism on integral homology for each i, $0 \leq i \leq \min(n,k)$.

The proof is an application of Smith theory, Mayer-Vietoris sequences, and the Comparison Theorem for spectral sequences. Smith theory comes into play via the following lemma (which we essentially proved in section I.A.2).

Lemma 2: Let $G(n) = U(n)$ or $Sp(n)$ and let $F: M \to M'$ be a stratified map of k-axial $G(n)$ manifolds. Further suppose that F induces an isomorphism on homology. Then for each i, the restriction of F to the fixed set of $G(n-i)$, $F^{G(n-i)}: M^{G(n-i)} \to M'^{G(n-i)}$ is also a homology isomorphism.

Proof: Let T be a maximal torus of $G(n-i)$. We first claim that $M^{G(n-i)} = M^T$. It clearly suffices to check this on each orbit. But each orbit is isomorphic to an orbit in the linear model, and in this case, $M(n,k)^{G(n-i)} = M(i,k) = M(n,k)^T$. (This argument is false for $O(n)$.)

Next, let $C(F)$ be the mapping cone of F. By hypothesis $C(F)$ is acyclic. $G(n)$ acts on $C(F)$. Since F is stratified the fixed set of the mapping cone is just the mapping cone of the map on the fixed sets, i.e.,

$$C(F)^{G(n-i)} = C(F^{G(n-i)}).$$

On the other hand, by the assertion in the first paragraph,

$$C(F)^{G(n-i)} = C(F)^{T} \quad .$$

By P. A. Smith's Theorem, $C(F)^T$ is acyclic. Therefore, $F^{G(n-i)}$ is a homology isomorphism. □

<u>Proof of the Main Lemma</u>: First suppose that $F_*: H_*(M) \to H_*(M')$ is a isomorphism. We wish to show that for each i, $(f|_{B_i})_*: H_*(B_i) \to H_*(B_i')$ is also an isomorphism. The proof is by induction on i. For $i = 0$, $B_0 = M_0 = M^{G(n)}$ and $f|_{B_0} = F^{G(n)}$, so this case follows from the previous lemma. Now suppose, by induction, that $f|_{B_j}$ is a homology isomorphism for each $j < i$. We make note of the following principle. If $P \to B_j$ and $P' \to B_j'$ are orientable fiber bundles (with the same fiber) and $g: P \to P'$ is a bundle map covering $f|_{B_j}$, then by the Comparison Theorem, g is also a homology isomorphism. Consider the orbit bundle $M_i \to B_i$. The associated principal bundle has total space $\{M_i\}^H$, where $H = G(n-i)$. But $\{M_i\}^H$ can be identified with the complement of tubular neighborhoods of the lower strata in M^H, that is

$$\{M_i\}^H = M^H - \cup\{N_j\}^H$$

where $\{N_j\}^H$ is the normal bundle of $\{M_j\}^H$ in M^H. Since F is stratified it can be altered by an equivariant isotopy so that $F|_{N_j}$ is a bundle map. Hence, we may assume that F^H preserves all the pieces in this decomposition of $\{M_i\}^H$. Hence, we may assume that F^H preserves all the pieces in this decomposition of $\{M_i\}^H$. By the Comparison Theorem and the inductive hypothesis, $F^H\colon \{N_j\}^H \to \{N_j'\}^H$ is a homology isomorphism. Moreover, if $j < \ell < i$ then $\{N_j\}^H \cap \{N_\ell\}^H$ is a bundle over B_j and the restriction of F^H to this piece is also a bundle map. So, by the above principle, F^H is also a homology isomorphism on all intersections. (Notice that the structure group of every bundle involved in this argument is connected, so there is problem with twisted coefficients.) Thus, a simple Mayer Vietoris argument shows that $F_i^H\colon \{M_i\}^H \to \{M_i'\}^H$ is a homology isomorphism. Finally we use the Comparison Theorem again to show that this map is a homology isomorphism on the base space, i.e., that $(f|_{B_i})_*$ is an isomorphism.

The proof in the other direction is similar. We must show that if each $(f|_{B_i})_*$ is an isomorphism, then so is F_*. By the Comparison Theorem, again, F is a homology isomorphism on each N_i as well as on their intersections. So, essentially the same argument with Mayer Vietoris sequences shows that F is a homology isomorphism on $M = \cup N_i$. □

2. The $O(n)$ case: First let us see what Smith theory tells us in the case of k-axial $O(n)$-actions. There is the following analogue of Lemma 2.

Lemma 3: Let $F\colon M \to M'$ be a stratified map of k-axial $O(n)$-manifolds. Suppose that F is a homology isomorphism with coefficients in $Z/2$.

Then for each i, $0 \leq i \leq n$, $F^{O(n-i)}: M^{O(n-i)} \to M'^{O(n-i)}$ *is also a homology isomorphism with coefficients in* $Z/2$.

Proof: The proof is the same as Lemma 2; however, we use a maximal $Z/2$-torus of $O(n-i)$ (consisting of the diagonal matrices in $O(n-i)$) rather than the maximal torus. □

For many of our purposes $Z/2$ coefficients are not good enough. We need the following improvement.

Lemma 4: *Let* $F: M \to M'$ *be a stratified map of* k-*axial* $O(n)$-*manifolds*. *Suppose that* F *is a homology isomorphism with coefficients in* Z. *Then for each* i *such that* $n - i$ *is even*, $F^{O(n-i)}$ *is also a homology isomorphism with coefficients in* Z.

Proof: When $n - i$ is even we can repeat the argument of Lemma 2. For if T is a maximal torus of $O(n-i)$, then $M^T = M^{O(n-i)}$, as before. (However, if $n - i$ were odd, then a maximal torus T of $O(n-i)$ would also be a maximal torus of $O(n-i-1)$. Hence, in this case $M^T = M^{O(n-i-1)} \neq M^{O(n-i)}$.) □

Remark: This lemma cannot be improved. For consider the Brieskorn sphere $\Sigma^{2n+1}(k,2,\ldots,2)$ where n is even. This supports a biaxial $O(n)$-action with fixed point set the circle $\Sigma^1(k,2)$. Moreover, the fixed point set of $O(n-i)$ is $\Sigma^{2i+1}(k,2,2,\ldots,2)$. Let W^{2n+1} be the complement of a disk about a fixed point in $\Sigma^{2n+1}(k,2,\ldots,2)$ and let D^{2n+1} be another small linear disk centered about a fixed point in W^{2n+1}. Then W^{2n+1} is a disk and the inclusion $D^{2n+1} \quad W^{2n+1}$ is a homology isomorphism. However, for i odd, W^{2i+1} has $H_i(W^{2i+1}) = Z/k$.

Hence, the induced map on the fixed point set of $O(n-i)$, $D^{2i+1} \hookrightarrow W^{2i+1}$ is not a homology isomorphism.

By using Lemma 3 instead of Lemma 2, the proof of the Main Lemma in the unitary and symplectic cases yields the following.

Weak Version of Main Lemma 5 ($O(n)$-case): Let $F: M \to M'$ be a stratified map of k-axial $O(n)$-manifolds covering $f: B \to B'$. Then $F_*: H_*(M;Z/2) \to H_*(M';Z/2)$ is an isomorphism if and only if for each i, $0 \leq i \leq \min(n,k)$, $(f|_{B_i})_*: H_*(B_i;Z/2) \to H_*(B_i';Z/2)$ is an isomorphism.

What we actually want, however, is to use Lemma 4 to prove a version of this lemma with integral coefficients. In order to state this analogue of the Main Lemma in the $O(n)$ case, we first need the notion of the "double branched cover of $B_i \cup B_{i-1}$ along B_{i-1}."

Let M be a k-axial $O(n)$-manifold. Let $O(i) \times O(n-i) \subset O(n)$ be the standard embedding. Then $O(i)$ acts on $M^{O(n-i)}$ and, in fact, $M^{O(n-i)}$ is a k-axial $O(i)$-manifold. (To see that the action is k-axial consider the linear model.) Let $H(i;M)$ be the union of the top two strata of $M^{O(n-i)}$, i.e., let

$$H(i;M) = \{M_i \cup M_{i-1}\}^{O(n-i)}.$$

Then $H(i;M)$ is clearly a smooth $O(i)$-manifold. Moreover, $H(i;M)/O(i) = B_i \cup B_{i-1}$. Consider the action of $SO(i)$ on $H(i;M)$. Since $O(i)$ acts freely on $\{M_i\}^{O(n-i)}$, $SO(i)$ also operates freely on this part. The isotropy group of $O(i)$ at $x \in \{M_{i-1}\}^{O(n-i)}$ is conjugate to $O(1)$. Since $SO(i) \cap O(1) = \{1\}$, $SO(i)$ also operates freely on this part. Hence, define

$$D_i(M) = H(i;M)/SO(i).$$

Then $D_i(M)$ is a smooth manifold supporting an action of $Z/2 = O(i)/SO(i)$. Clearly, $D_i(M)^{Z/2} = B_{i-1}$ and $D_i(M)/(Z/2) = B_i \cup B_{i-1}$. Thus, it makes sense to call $D_i(M)$ the <u>double branched cover of</u> $B_i \cup B_{i-1}$ <u>along</u> B_{i-1}. By convention, set $D_0(M) = B_0$ and $D_{k+1}(M) = B_k$. Notice that a stratified map $F: M \to M'$ induces a $Z/2$-equivariant map $D_i(F): D_i(M) \to D_i(M')$. The strong version of the Main Lemma in the $O(n)$ case is the following.

<u>Main Lemma 6</u> ($O(n)$-case): <u>Let</u> $F: M \to M'$ <u>be a stratified map of</u> k-<u>axial</u> $O(n)$-<u>manifolds. Then</u> F <u>is an integral homology isomorphism if and only if for each</u> i, $0 \leq i \leq k+1$, <u>such that</u> $n - i$ <u>is even</u>, $D_i(F): D_i(M) \to D_i(M')$ <u>is an integral homology isomorphism</u>.

The basic idea in proving this lemma is the same as in the unitary and symplectic case only now we want to take M apart two strata at a time. However, the technicalities involved make the proof an order of magnitude more difficult and we will postpone the proof to Appendix 2.

<u>Remark</u>: If $f: X \to Y$ is an equivariant map of $Z/2$-manifolds and f is an integral homology equivalence, then Smith theory tells us that the maps

$$X^{Z/2} \longrightarrow Y^{Z/2}$$

$$X - X^{Z/2} \longrightarrow Y - Y^{Z/2}$$

$$(X - X^{Z/2})/(Z/2) \longrightarrow (Y - Y^{Z/2})/(Z/2)$$

are all homology equivalence with coefficients in Z/2. There is one case in which we can say more--when the fixed point sets are of codimension one. In this case all of the above maps are integral homology equivalences. The proof of this was essentially given in section I.A.2 (see [5]). In particular, in regard to Main Lemma 6, if $D_k(M) \to D_k(M')$ is an integral homology equivalence then so are $B_k \to B'_k$ and $B_{k-1} \to B'_{k-1}$. Since the map is always a homology equivalence on either $D_k(M)$ or $D_{k+1}(M) = B_k$, we conclude that if $F: M \to M'$ is a homology equivalence, then so is the induced map on the top stratum $f|_{B_k} : B_k \to B'_k$.

B. <u>An Application of Borel's Formula</u>

A <u>linear</u> k-<u>axial</u> G(n)-<u>action on a sphere</u> is a linear action of G(n) on $S^{dkn+m-1}$, where $S^{dkn+m-1}$ is the unit sphere in $M(n,k) \times R^m$ (and where d = 1,2 or 4 as G(n) = O(n), U(n) or Sp(n)). Suppose that G(n) acts k-axially on a homology sphere Σ. If the fixed point set is nonempty, then by Smith theory it is also homology sphere of some dimension, say, m - 1. Since the normal representation at a fixed point is equivalent to M(n,k) it follows that

$$\dim \Sigma = \dim M(n,k) + \dim \Sigma_0$$

$$= dkn + m - 1$$

and that each stratum of Σ has the same dimension as the corresponding stratum of $S^{dkn+m-1}$.

However, it is not clear, a priori, that the same phenomenum persists if the fixed point set is empty. The only linear action with empty fixed point set is the action on S^{dkn-1} (the unit sphere in $M(n,k)$). If $G(n)$ acts k-axially on a homology sphere Σ and if the fixed point set is empty, then the natural conjecture is that $\dim \Sigma = dkn - 1$, that $\Sigma_i \neq \emptyset$ for $i > 0$, and that each Σ_i has the same dimension as the i-stratum of S^{dkn-1}. In fact this is true and it was proved by the Hsiangs in [2] by using Borel's formula (see [1]).

Theorem 7 (Borel's Formula): *Suppose G is a torus acting on an integral homology sphere Σ^s of dimension s. If $H \subset G$ is a subtorus then by Smith theory, Σ^H is a homology sphere. Let $r(h) = \dim \Sigma^H$. (If Σ^H is empty set $r(H) = -1$.) Then*

$$s - r(G) = \sum_H r(H) - r(G),$$

where H runs through all toral subgroups of G of codimension one. This identity remains valid if Σ is a mod p homology sphere, G is an elementary abelian p-group (a "Z/p-torus") and H is a subgroup of index p.

Theorem 8 (the Hsiangs): *Let $G(n)$, $n > 1$, act k-axially on a homology sphere Σ^s of dimension s, and let $r = \dim \Sigma_0$ (set $r = -1$ if $\Sigma_0 = \emptyset$). Then*

$$s = dkn + r.$$

Furthermore, Σ_i has the same dimension as the i-stratum of S^{dkn+r}. In particular, $\Sigma_i \neq \emptyset$ for $i > 0$.

Proof: If $G(n) = U(n)$ or $Sp(n)$ let $T \subset G(n)$ be the standard maximal torus. If $G(n) = O(n)$, let T be the maximal Z/2-torus. Let $x \in \Sigma$

and consider $T_x = T \cap G(n)_x$. We see that T_x is corank one if and only if $G(n)_x = G(n)_{e_i}$, where $G(n)_{e_i}$ is the isotropy group of the standard representation at a standard basis vector $e_i \in \mathbb{F}^n$. ($G(n)_{e_i}$ is conjugate to $G(n-1)$.) Thus, T_x is corank one if and only if $T_x = T_i$, where $T_i = T \cap G(n)_{e_i}$ (T_i consists of all diagonal matrices with 1 in the ith position). Borel's formula yields

$$s - r = \sum_{i=1}^{i=n} (r(T_i) - r).$$

Since the T_i's are conjugate in $G(n)$, $r(T_1) = r(T_2) = \ldots = r(T_n)$. Thus,

$$s - r = n(r(T_1) - r).$$

This implies that $\Sigma^{G(n-1)} \neq \emptyset$. For otherwise $r = -1 = r(T_1)$, in which case $s = -1$.

Next, let $x \in \Sigma^{G(n-1)} = \Sigma^{T_1}$, and let τ be the tangent space of Σ at x. By the Slice Theorem, $\dim \Sigma^{G(n-1)} = \dim \tau^{G(n-1)}$. The normal representation at x is $M(n-1),k-1)$. The representation of $G(n-1)$ on the tangent space to the orbit at $x = T(S^{dn-1})$ is equivalent to $1\rho_{n-1} + (d-1)1$ on $M(n-1,1) + R^{d-1}$. Hence, the part of τ which is acted on nontrivially by $G(n-1)$ is equivalent to $M(n-1),k-1) + M(n-1,1) \cong M(n-1,k)$. It follows that

$$r(T_1) = \dim \tau^{G(n-1)}$$

$$= s - dk(n-1).$$

Substituting this in Borel's formula

$$(s-r) = n(s-dk(n-1)-r)$$

$$(n-1)(s-r) = ndk(n-1)$$

$$s - r = dkn .$$

To prove the last two sentences in the theorem, let $S = S^{dkn+r}$. Our calculation of the dimension of $\Sigma^{G(n-1)}$ shows that $\dim \Sigma^{G(n-1)} = \dim S^{G(n-1)}$ and hence, that $\dim \Sigma_1 = \dim S_1$. Since the dimension of Σ_i is the same as the dimension of the i-stratum of the normal bundle of Σ_1, it follows that $\dim \Sigma_i = \dim S_i$. □

Remark: Actually this is only a small part of the theorem proved by the Hsiangs in pages 744-750 of [2]. They use Borel's Formula and representation theory to show that if $G^d(n)$ acts smoothly on a homotopy sphere and if the principal orbit type is a Stiefel manifold $G^d(n)/G^d(n-k)$, $k < n$, then (under some dimension restrictions) the action must be k-axial. These dimension restrictions were later removed in [3]. Thus, they proved that *if the principal orbit type of* $G(n)$ *on* Σ *is a Stiefel manifold, then the action is automatically multiaxial*. The proof of this used Wu-yi Hsiang's theory of "geometric weight systems" which is a systematic generalization of the ideas behind Borel's Formula.

C. *The bundle of principal orbits of a* k-*axial action on a sphere is trivial*, if $n \geq k$.

Proposition 9: *Suppose that* $G(n)$ *acts* k-*axially on a homology sphere* Σ *and that* $n \geq k$. *Then* B_k, *the top stratum of the orbit space, is acyclic over the integers*.

Proof: Suppose that the dimension of the fixed point set is $m - 1$. Then $\dim \Sigma = dkn + m - 1$. We will prove the proposition except in the case where $m = 0$ and $n = k$.

Case 1, $m > 0$: In this case the fixed point set is a (nonempty) homology sphere. Hence, it contains at least two points, say x and y. Let S be a slice at x, i.e., let S be an invariant Euclidean neighborhood of x such that the action on S is equivalent to the linear action on $M(n,k) \times R^{m-1}$. Then the inclusion $S \subset \Sigma - \{y\}$ is a homology equivalence. Let $L = S/G(n)$. Then the top stratum of L is contractible. Hence, by Lemmas 1 and 6 and the remark at the end of section A.2, the top stratum of $(\Sigma - \{y\})/G(n)$ is acyclic. But the top stratum of $\Sigma - \{y\}$ is equal to the top stratum of Σ. Hence, B_k is acyclic.

Case 2, $m = 0$ and $n > k$: To prove this case, we restrict the action to $G(n-1)$. Let $A = \Sigma/G(n-1)$. The dimension of Σ is $dk(n-1) + dk - 1$; hence, by the result of the previous section, A_0 is a nonempty homology sphere (of dimension $dk - 1$). Since $n - 1 \geq k$, Case 1 applies, so A_k is acyclic. Since A is a manifold with boundary with interior A_k, this implies that A is acyclic. Consider the natural surjection $p: A \to B$, which sends a $G(n-1)$ orbit to its orbit under $G(n)$. We propose to show that B (and hence B_k) is acyclic by showing that point inverse images under p are acyclic (in fact, they are disks). A $G(n)$-orbit is a Stiefel manifold $V_{n-i} = G(n)/G(n-i)$. The inverse image of this orbit in A is isomorphic to $V_{n,i}/G(n-1)$. Hence, our assertion follows from the following lemma.

Lemma 10: Let $G(n-1) \subset G(n)$ be the standard inclusion and let $\psi: G(n-1) \times V_{n,i} \to V_{n,i}$ be the restriction of the canonical transitive $G(n)$-action to $G(n-1)$. Then $V_{n,i}/G(n-1) = D^{di}$.

Proof: Let $q: \mathbb{E}^n \to \mathbb{E}$ be projection on the first factor and let $(x_1,\ldots,x_i)$ be an i-frame. Define $f: V_{n-i} \to \mathbb{E}^i$ by $f(x_1,\ldots,x_i) = (q(x_1),\ldots,q(x_i))$. The image of f is clearly the unit disk D^{di}. If $g \in G(n-1)$, then $f(gx_1,\ldots,gx_i) = f(x_1,\ldots,x_i)$; hence f induces a map $\bar{f}: V_{n,i}/G(n) \to D^{di}$. We claim that $\bar{f}$ is injective and therefore, a homeomorphism. To see this suppose that $f(x_1,\ldots,x_i) = f(y_1,\ldots,y_i)$. If $X = (X_1,\ldots,X_n) \in \mathbb{E}^n$ set $X' = (X_2,\ldots,X_n) \in \mathbb{E}^{n-1}$. Since x_j and x_ℓ are orthogonal vectors

$$\begin{aligned} 0 &= x_j \cdot x_\ell \\ &= x_j' \cdot x_\ell' + q(x_j)q(x_\ell). \end{aligned}$$

Thus,

$$\begin{aligned} x_j' \cdot x_\ell' &= -q(x_j)q(x_\ell) \\ &= -q(y_j)q(y_\ell) \\ &= y_j' \cdot y_\ell'. \end{aligned}$$

If two i-tuples of vectors in $\mathbb{E}^{n-1}$ $(x_1',\ldots,x_i')$ and $(y_1',\ldots,y_i')$ have all dot products equal, then they differ by a unitary transformation, i.e., there is $g \in G(n-1)$ such that $(gx_1',\ldots,gx_i') = (y_1',\ldots,y_i')$. In other words, $(x_1,\ldots,x_i)$ and $(y_1,\ldots,y_i)$ lie in the same orbit under $G(n-1)$. Thus, $\bar{f}$ is injective and therefore, a homeomorphism. □

This completes the proof of the Proposition except in the case $m = 0$ and $n = k$. The analysis in this case is similar, but more subtle and we omit it.

Corollary 11: Suppose $G(n)$ acts k-axially on an integral homology sphere Σ and that $n \geq k$. Then the k-orbit bundle $P_k \to B_k$ is a trivial $G(k)$-bundle.

Proof: B_k is acyclic by the above proposition. Thus $P_k \cong G(k) \times B_k$ by obstruction theory.

□

Therefore, provided $n \geq k$, the Structure Theorem applies to k-axial $G(n)$-actions on homology spheres. In particular, there is a stratified map $\Sigma \to M(n,k)$.

Now suppose that a closed oriented manifold M supports a k-axial $G(n)$-action with $n \geq k$. Further suppose that the k-orbit bundle of M is trivial. If M_0 is empty, then the stratified map $M \to M(n,k)$ has image contained in $M(n,k) - \{0\}$. Hence we can assume that the image is contained in the unit sphere $S^{dkn-1} \subset M(n,k)$. Otherwise, let $\dim M_0 = m - 1$. Consider the linear action on $S^{dkn+m-1}$. Let L be its orbit space. Notice that except for the fixed point set each stratum of $S^{dkn+m-1}$ is homotopy equivalent to the corresponding stratum of $M(n,k)$. In fact, the i-stratum of $S^{dkn+m-1}$ is equivalent to $M(n,k)_i \times R^{m-1}$. Choose any map $F_0: M_0 \to S^{m-1}$. The above observation together with the proof of the Structure Theorem shows that F_0 can be extended to a stratified map $F: M \to S^{dkn+m-1}$. As in the Structure Theorem, such an F is unique up to stratified homotopy, and a choice of trivialization of the bundle of principal orbits, only now we are free to choose F_0. In particular, we can choose F_0 to be of degree one. I claim that if F_0

is of degree one, then so is F. For let N_0 and N_0' be the normal bundles of M_0 in M and of S^{m-1} in $S^{dkn+m-1}$, respectively. Then F induces a map of Thom spaces $T(N_0) \to T(N_0')$ which is an isomorphism on the top homology class. Hence, consideration of the commutative diagram

$$\begin{array}{ccc} M & \xrightarrow{\quad F \quad} & S^{dkn+m-1} \\ \text{collapse} \big\downarrow & & \big\downarrow \text{collapse} \\ T(N_0) & \longrightarrow & T(N_0') \end{array}$$

shows that F is of degree one. These remarks lead us to the following.

Proposition 12: *Suppose that* $n \geq k$ *and that* $G(n)$ *acts* k-*axially on an integral homology sphere* Σ *of dimension* $dkn + m - 1$. *Then there exists a stratified map* $F\colon \Sigma \to S^{dkn+m-1}$. *Furthermore,* F *can be chosen to be of degree* ± 1 *except possibly in the case where* $G(n) = O(n)$, $m = 0$, k *is even, and* n *is odd*.

Proof: For $m > 0$, the Proposition follows immediately from the preceding remarks. If $m = 0$, the map $F\colon \Sigma \to S^{dkn-1}$, given by the Structure Theorem, is unique up to a stratified homotopy and choice of trivialization of the k-orbit bundle. Hence, its degree is predetermined (at least up to sign). Let $f\colon B \to L$ be the induced map of orbit spaces. Consider the restriction of f to the 1-stratum. First notice that $L_1 = \mathbb{F}P^{k-1}$ (where $\mathbb{F} = \mathbb{R}$, $\mathbb{C}$ or $\mathbb{H}$ as $G(n) = O(n)$, $U(n)$ or $Sp(n)$).

If $G(n) = U(n)$ or $Sp(n)$, then $\Sigma^{G(n-1)} \underset{Z}{\sim} S^{dk-1}$ by Smith theory and therefore, $B_1 \underset{Z}{\sim} \mathbb{F}P^{k-1}$. On the other hand, the $G(1)$-bundle

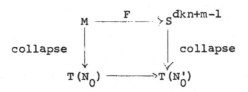

$\Sigma^{G(n-1)} \to B_1$ is just the pullback of $S^{dk-1} \to \mathbb{F}P^{k-1}$ via f. It follows that f must pull back the Euler class of this bundle to a generator of $H^d(B_1;Z)$. Hence, $(f|_{B_1})_*: H_*(B_1) \to H_*(L_1)$ and $(F^{G(n-1)})_*: H_*(\Sigma^{G(n-1)}) \to H_*(S^{dk-1})$ are isomorphisms. A simple argument using the Thom space of the normal bundle of $\Sigma^{G(n-1)}$, similar to the one given above, now shows that F has degree ± 1.

Next consider the case $G(n) = O(n)$. If n is even, we can essentially use the above argument. For by Smith theory $\Sigma^{O(n-2)} \underset{Z}{\sim} S^{2k-1}$, and therefore, $\Sigma^{O(n-2)}/SO(2) = D_1(\Sigma)$ has the integral cohomology of $\mathbb{C}P^{k-1}$. So as above, $\Sigma^{O(n-2)} \to S^{2k-1}$ is of degree one and then by the Thom space argument F is of degree one.

If n is odd and k is odd, we can use a different argument. Again, Smith theory yields $\Sigma^{O(n-1)} \underset{Z}{\sim} S^{k-1}$ and $B_1 \underset{Z[Z_2]}{\sim} \mathbb{R}P^{k-1}$. Also, $D_1(\Sigma) \underset{Z/2}{\sim} \mathbb{C}P^{k-1}$. But this implies that $D_1(\Sigma)$ is a rational homology complex projective space. Therefore, we can calculate the G-signature of the involution on $D_1(\Sigma)$. This involution on $D_1(\Sigma)$ has fixed point set B_1. Let ν be the normal bundle of B_1 in $D_1(\Sigma)$. This has a twisted Euler class $\chi(\nu) \in H^{k-1}(\mathbb{R}P^{k-1};Z_-)$. Since k is odd, $H^{k-1}(\mathbb{R}P^{k-1};Z_-) = Z$. (Here Z_- denotes twisted coefficients.) Since $D_1(F)$ is stratified, f must pull back the linear twisted Euler class to $\chi(\nu)$. On the other hand, we can calculate $\chi(\nu)$ by using the G-signature Theorem (see page 188 in [4]). In fact, $\chi(\nu)$ is a generator of $H^{k-1}(\mathbb{R}P^{k-1},Z_-)$. It follows that $f|_{B_1}$ has twisted degree ± 1, from which we deduce that $F^{O(n-1)}: \Sigma^{O(n-1)} \to S^{k-1}$ and $F: \Sigma \to S^{kn-1}$ are both of degree ± 1. □

<u>Remark</u>: When n is odd consider the Brieskorn sphere $\Sigma^{2n-1}(\ell,2,\ldots,2)$.

The classifying map $\Sigma^{2n-1}(\ell,2,\ldots,2) \to S^{2n-1}$ has degree ℓ. Suppose we can write $n = tn'$. Consider the embedding $t\rho_{n'}: O(n') \to O(n)$. Restricting the $O(n)$-action on $\Sigma^{2n-1}(\ell,2,\ldots,2)$ to $O(n')$ gives a fixed point free 2t-axial $O(n')$-action. The classifying map $\Sigma^{2n-1}(\ell,2,\ldots,2) \to S^{2tn'-1}$ again has degree ℓ. Notice that in these examples t must be odd. Hence, for n odd and $k \equiv 2 \pmod 4$ there are k-axial $O(n)$ actions on homotopy sphere Σ^{kn-1} with the classifying map of any odd degree. For n odd and $k \equiv 0$ (4), one can show by using surgery theory and results in Chapter VI that the classifying map must have degree $\pm 1 \pmod 8$. We conjecture that any such degree can occur.

The next question is to decide if we can recognize when Σ is simply connected (and hence, a homotopy sphere) by looking at its orbit space. It turns out that the fundamental group of Σ is determined by the fundamental group of the top stratum of B.

Proposition 13: Suppose that M is a k-axial $G(n)$-manifold with $n \geq k$. If $G(n) = O(n)$ further suppose that $n \geq k + 2$. Let $B = M/G(n)$. Then M is simply connected $\Leftrightarrow$ B_k is simply connected.

Remark: In the case where $G(n) = O(n)$ and $n = k$ or $n = k + 1$, the proposition is still true provided we assume that M is a homology sphere. However, this requires a separate argument, which we shall not give here. A separate argument is also needed when $G(n) = U(n)$ and $n = k$.

Proof of the Proposition: ($\Rightarrow$) suppose M is 1-connected. Consider the bundle $M_k \to B_k$. The fiber is $G(n)/G(n-k)$, which is connected; hence, $\pi_1(M_k) \to \pi_1(B_k)$ is surjective. Next consider the inclusions

$M_k \subset M$ and $B_k \subset B$. Since B is a manifold with boundary with $\mathrm{int}(B) = B_k$, it follows that $\pi_1(B_k) \cong \pi_1(B)$. The diagram

$$\begin{array}{ccc} 0 = \pi_1(M) & \longrightarrow & \pi_1(B) \\ \uparrow & & \uparrow \wr\!\wr \\ \pi_1(M_k) & \xrightarrow{\text{onto}} & \pi_1(B_k) \end{array}$$

now shows that $\pi_1(B_k) = 0$.

($\Leftarrow$) Suppose $\pi_1(B_k) = 0$. Since $\pi_1(G(n)/G(n-k)) = 0$ (provided $G(n)/G(n-k) \neq O(n)/O(1)$, $O(n)$ or $U(n)$), it follows that $\pi_1(M_k) = 0$. From the hypotheses, it can be shown that the union of the lower strata has codimension greater than two (except when $G(n) = U(n)$ and $n = k$); and hence, that $\pi_1(M) = \pi_1(M_k)$. To see this, note that the $(k-1)$-stratum has normal representation $M(n-k+1,1)$; so that the union of the lower strata has codimension $d(n-k+1)$ and $d(n-k+1) > 2$. (When $G(n) = U(n)$ and $n = k$, then $d(n-k+1) = 2$, so this case requires a separate argument, which is left as an exercise.) □

In summary, we have the following theorem.

Theorem 14: *Suppose* $n \geq k$ *and that* $G(n)$ *acts* k-*axially on a closed manifold* M. *Suppose that the bundle of principal orbits of* M *is trivial and that the fixed point set is nonempty and of dimension* $m - 1$. *Then dim* $M = dkn + m - 1$ *and*

(1) *There exists a stratified map* $F: M \to S^{dkn+m-1}$ *of degree one*. *Suppose* F *covers the map* $f: B \to L$ *of orbit spaces. Then* M *is a homology sphere if and only if* F_* *is an isomorphism, i.e., if and only if*

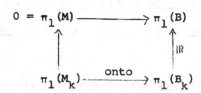

(2) For $G(n) = U(n)$ or $Sp(n)$, $f|_{B_i}$ is a homology isomorphism for each i, $0 \leq i \leq k$.

(3) For $G(n) = O(n)$, then $D_i(F): D_i(M) \to D_i(S)$ is a homology isomorphism for each i, $0 \leq i \leq k+1$, such that $n - i$ is even.

Moreover, M is a homotopy sphere if and only if, in addition, B is simply connected.

References

[1] A. Borel et al., Seminar on Transformation Groups, Ann. of Math. Studies 46, Princeton University Press (1961).

[2] W. C. Hsiang and W. Y. Hsiang, Differentiable actions of compact connected classical groups: I, Amer. J. Math. 89 (1967), 705-786.

[3] ______________________________, Differentiable actions of compact Connected Lie groups: III, Ann. of Math. 99 (1974), 220-256.

[4] C. T. C. Wall, Surgery on Compact Manifolds, Academic Press, New York, 1970.

[5] C. T. Yang, Transformation groups on a homological manifold. Trans. Amer. Math. Soc. 87 (1958), 261-283.

VI. CONCORDANCE CLASSES OF ACTIONS ON SPHERES

Let us review the situation. Suppose that $n \geq k$ and that $G(n)$ acts k-axially on a homotopy sphere Σ. Let B be its orbit space. In the previous lecture we showed that Σ is the pullback of the linear model $M(n,k)$ via some map $f: B \to H_+(k)$. Furthermore, we showed that B is simply connected and that it is "homologically modeled" on the orbit space of the linear action on the sphere of the same dimension as Σ. Moreover, by the Classification Theorem of Lecture 3, k-axial $G(n)$-manifolds over B are classified by the set of stratified homotopy classes of maps from B to $H_+(k)$. (B_k is contractible, so for $G(n) = U(n)$ or $Sp(n)$ there is, up to homotopy, only one trivialization of $P_k \to B_k$; however, for $G(n) = O(n)$, we must divide $\pi_0 \mathrm{Hom}(B;H_+(k))$ by the action of $Z/2 = [B_k;O(k)]$.). Also, at least in the $U(n)$ and $Sp(n)$ cases, we can recognize when the pullback $f^*M(n,k)$ is a homotopy sphere. For in these cases, necessary and sufficient conditions are that B_0 is an integral homology sphere and that $f|_{B_i}: B_i \to H_+(k)_i$ induces an isomorphism in integral homology for all $i > 0$.

Although this is in some sense a complete classification of such actions on homotopy spheres, it is unsatisfactory in several ways. First of all, we don't yet have a good method of constructing such G's together with stratified maps $f: B \to H_+(k)$. Secondly, the calculation of the set of stratified homotopy classes of maps from B to $H_+(k)$ is a problem of considerable difficulty. Thirdly, since the fundamental groups of the lower strata do not have to be correct; there are some trivial ways to modify the orbit space of a linear action to introduce more fundamental group into the lower strata (similar to the construction of Montgomery and Samelson in Chapter 1) and we would like to ignore these trivial modifications. Finally, we would like to be able to decide the differential structure on Σ from B and from f. All these problems are solved by introducing the notion of concordance.

Roughly speaking, two k-axial $G(n)$-actions $\varphi: G(n) \times M \to M$ and $\varphi': G(n) \times M' \to M'$ are <u>concordant</u> if there is a k-axial $G(n)$-action $\psi: G(n) \times M \times [0,1] \to M \times [0,1]$ such that $\psi(G(n) \times M \times \{i\}) = \dot{M} \times \{i\}$, $i = 0,1$, and such that the restriction of ψ to $M \times \{0\}$ is equivalent to (M,φ) while its restriction to $M \times \{1\}$ is equivalent to (M',φ').

Actually in order to make concordance classes of actions on homotopy spheres into a group we need some orientation conventions. For $G(n) = U(n)$ or $Sp(n)$, it is only necessary to modify the above definition by requiring that are manifolds be oriented, that "equivalence" be interpreted as "orientation preserving equivariant diffeomorphism", and that $(M \times \{0\}, \psi|_{M \times \{0\}})$ be equivalent to (M,φ), while $(M \times \{1\}, \psi|_{M \times \{1\}})$ be equivalent to $-(M',\varphi')$. We could try to require other strata to be oriented; however, for $U(n)$ and $Sp(n)$, it is not hard to see that an orientation for the total space induces an orientation for each stratum.

For $G(n) = O(n)$, this is not true. It turns out however, that it is only necessary to specify an orientation for M and one for $M^{O(1)}$. Thus, an (M,φ) is <u>oriented</u> if we specify both of these orientations. (These two orientations determine orientations for all the double branched covers $D_i(M)$.) Also, for $G(n) = O(n)$, "equivalence" should be interpreted as an equivariant diffeomorphism preserving both orientations. To quote Bredon [3], "we will not attempt the difficult and pointless task of tracing these conventions through our constructions."

If M and M' are two oriented k-axial $G(n)$-manifolds of the same dimension with nonempty fixed point sets, then it is an easy exercise to define the equivariant connected sum $M \# M'$, also an oriented k-axial $G(n)$-manifold.

Now we come to our main object of study. Let

$$\theta^d(k,n,m)$$

denote the set of concordance classes of k-axial $G^d(n)$-actions on homotopy sphere of dimension $dkn + m - 1$. (It follows that $m - 1$ is the dimension of the fixed point set.)

<u>Theorem 1</u>: <u>The set</u> $\theta^d(k,n,m)$ <u>is an abelian group under connected sum. A</u> k-<u>axial action</u> φ: $G^d(n) \times \Sigma \to \Sigma$ <u>on a homotopy sphere</u> Σ <u>represents the zero element of this group if and only if</u> φ <u>extends to a</u> k-<u>axial action on a disk. The inverse is given by reversal of orientation</u> (<u>or both orientations when</u> $d = 1$).

<u>Proof</u>: The standard arguments work. See, for example, [7] or page 339 in [1].

A. <u>The plan of attack</u>.

Suppose $G^d(n)$ acts k-axially on a homotopy sphere $\Sigma^{dkn+m-1}$ and that $n \geq k$. Let D be the unit disk in $M^d(n,k) \times R^m$ and let $S = \partial D$. In the previous chapter it was shown that there is a degree one stratified map $F: \Sigma \to S$. (It follows that F is a homotopy equivalence.) In section B we will prove that Σ bounds a parallelizable k-axial G(n)-manifold V such that F can be extended to a stratified map of pairs F: $(V,\Sigma) \to (D,S)$.

Let f: $(A,B) \to (K,L)$ be the map of orbit spaces induced by F. First suppose that G(n) = U(n) or Sp(n). It follows from the Main Lemma 1 and Proposition 13 of the previous chapter that F is a homotopy equivalence on the boundary if and only if $\pi_1(B) = 0$ and $f|_{B_i}: B_i \to L_i$ is an integral homology equivalence for each $i \geq 0$. Moreover, V will be a disk precisely when $\pi_1(A) = 0$ and when $f|_{A_i}$ is a homology equivalence for all $i \geq 0$ (provided dim $V \geq 5$).

Therefore, we try to do "surgery" on f rel B to get a new orbit space A' together with a new map f': $A' \to K$ so that $\pi_1(A') = 0$ and so that $f'|_{A_i'}$ is a homology equivalence. If we succeed then $V' = f'^*(D)$ will be a contractible manifold with $\partial V \cong \Sigma$. Hence, if this surgery is possible, Σ represents zero in $\theta^d(k,n,m)$. The converse is also true. In [5], W. C. Hsiang and I showed that for U(n) and Sp(n), the only obstruction to doing this surgery is the index or Kervaire invariant of the fixed point set $V_0 = A_0$. Furthermore, if k is odd, this obstruction automatically vanishes. On the other hand, it will follow from the Realization Theorem for surgery obstructions that for k even, any index (divisible by 8) or Kervaire invariant can occur. Alternatively, for k even, it can be seen that all possible nonzero obstructions are realized by actions on Brieskorn spheres.

The situation for G(n) = O(n) is analogous. Except that now we only know at the outset that $\pi_1(B) = 0$ and that for each i, with $i \equiv n \pmod 2$, $D_i(F): D_i(\Sigma) \to D_i(S)$ is a homology equivalence. (This implies that $f|_{B_i}$ is a $Z_{(2)}$ homology equivalence.) Similar conditions must be satisfied in order to construct a contractible V'. One approach would be to try to do surgery rel B on f to a $Z_{(2)}$-homology equivalence on each stratum. If this is possible we obtain V', acyclic over $Z_{(2)}$. Thus such a surgery theory leads to calculation of the group of k-axial O(n) actions on Z/2-homology spheres (with trivial bundle of principal orbits) modulo actions on "Z/2-homology h-cobordisms." However, to calculate actual concordance classes we are forced to do Z/2 equivariant surgery on the "double branched covers", $D_i(V)$. The calculation of these surgery obstructions in the O(n) case is an order of magnitude more difficult than before. It will appear as joint work with Wu-chung Hsiang and John Morgan [6].

By "surgery", we essentially mean surgery on a stratified space (as in [4]). This type of surgery is a generalization of surgery on a manifold with boundary (which has two strata). The way in which the surgery obstructions are computed can also be illustrated by consideration of this example. So suppose that $\varphi: (M,\partial M) \to (N,\partial N)$ is a normal map and also suppose for simplicity that $\pi_1(N) = \pi_1(\partial N) = 0$. If the boundary is nonempty, it is well-known that we can complete surgery to a homotopy equivalence of pairs. One way to see this is as follows. First we do surgery on $\varphi|_{\partial M}$. There is no obstruction (i.e., no index or Kervaire invariant), since $\varphi|_{\partial M}$ is a boundary (of φ). In this argument we are "looking up one stratum." One continues by trying to do surgery rel boundary on $M \cup X$ where X is the trace of the surgery on the boundary. Again, we might meet an obstruction. If so we simply change the cobordism X by adding the negative of this obstruction. Hence, surgery will again be possible. In this argument we are "going down one stratum and changing the cobordism."

The problem of doing surgery on orbit spaces is analogous. For $G(n) = U(n)$ or $Sp(n)$, we must solve a sequence of (ordinary) simply connected surgery problems indexed by $\{0,1,\ldots,k\}$. In these cases we use the fact that $\partial_i A_{i+1} \to A_i$ is a fiber bundle with fiber $\mathbb{C}P^{k-i-1}$ or $\mathbb{H}P^{k-i+1}$. If $k - i - 1$ is even, we can look up one stratum and conclude that the surgery obstruction at the ith stage vanishes. If $k - i - 1$ is odd the obstruction is indeterminacy (by going down one stratum and changing the cobirdism). One of these arguments always works except possibly in the case where k is even and $i = 0$. In this case we would like to go down one stratum, but there is no lower stratum. Hence, we are left with a legitimate obstruction (which can be realized by a Brieskorn example).

The $O(n)$ case is more complicated. The strata are not simply connected and half of the time they are nonorientable. It turns out, however, the entire obstruction is concentrated in the bottom two strata. Here the analysis involves sometimes looking up two strata and sometimes showing that the obstructions are indeterminacy by going down two strata. The final result will be stated more precisely in section C.

B. Multiaxial actions bound.

Suppose that B is a local orbit space. Since this means that we know the "smooth functions" on B for any $b \in B$, we can define the tangent space $T_b(B)$. In general the dimension of $T_b(B)$ will not be constant. However, as we noted in Lecture 3 for any $x \in H_+(k)$, $T_x(H_+(k)) = H(k)$. ($H(k)$ is the vector space of all $(k \times k)$ hermitian matrices.) It follows from this observation that if B is modeled on $H_+(k)$, then

$$T(B) = \bigcup_{b \in B} T_b(B)$$

is a vector bundle over B. The next proposition was observed by Bredon [3] in the case k = 2 and in fact, his proof works in general.

Proposition 2 (Bredon): *Suppose that M is a k-axial G(n)-manifold, $n \geq k$, with orbit space B. If the bundle of principal orbits is trivial, then the following four conditions are equivalent.**

(1) M *is stably parallelizable.*

(2) $T(B_k)$ *is trivial.*

(3) $T(B)$ *is trivial.*

(4) *As a G(n)-vector bundle,* $T(M)$ *is stably equivalent to the product bundle* $M \times M(n,k)$.

Proof: We will show that (1) ⇒ (2) ⇒ (3) ⇒ (4) ⇒ (1).

(1) ⇒ (2): Since M is stably parallelizable, $T(M_k)$ is stably trivial. But $M_k = G(n)/G(n-k) \times B_k$. Since Stiefel manifolds are stably parallelizable, it follows that so is B_k. But B_k is the interior of a manifold with nonempty boundary, hence, B_k is actually parallelizable.

(2) ⇒ (3): The inclusion $B_k \subset B$ is a homotopy equivalence. Hence, $T(B)$ is trivial ⇔ $TB_k = TB|_{B_k}$ is trivial.

(3) ⇒ (4): By the Structure Theorem M is equivalent to the

*Assuming the action has more than one nonempty stratum. Otherwise we would only have that $T(B_k)$ and $T(B)$ are stably trivial.

pullback of $M(n,k)$ via $f: B \to M(n,k)$. Thus, we may identify M with the subset $\psi^{-1}(0) \subset M(n,k) \times B$, where $\psi: M(n,k) \times B \to H(k)$ is defined by $\psi(x,b) = \pi(x) - f(b)$. It is not hard to see that $0 \in H(k)$ is a regular value of ψ (in fact, this is how the Pullback Theorem is proved). Therefore, the normal bundle of M in $M(n,k) \times B$, being the pullback of the normal bundle of $0 \in H(k)$, is trivial. It follows that up to stable equivalence of $G(n)$-vector bundles

$$T(M) \cong T(M(n,k) \times B)\big|_{\psi^{-1}(0)}$$

$$\cong M \times M(n,k).$$

(4) $\Rightarrow$ (1): Trivial. $\square$

<u>Theorem 3</u>: <u>Suppose that</u> $n \geq k$ <u>and that a</u> k-<u>axial</u> $G(n)$-<u>manifold</u> M <u>is a pullback of the linear model</u> $M(n,k)$. <u>Then there is a</u> k-<u>axial</u> $G(n)$-<u>manifold</u> V, <u>also a pullback of the linear model, with</u> $\partial V = M$. <u>Moreover, if</u> M <u>is stably parallelizable, then so is</u> V.

We shall give three different proofs.

<u>Proof 1</u>: Suppose that $M = f^*(M,(n,k))$ where $f: B \to H_+(k)$. Define a sequence of manifolds

$$M(k) \subset M(k+1) \subset \ldots M(m) \subset \ldots$$

where $M(m) = f^*(M(m,k))$, $m \geq k$. $M(m)$ is a k-axial $G(m)$-manifold over B. Notice that $M(n) = M$. Also, $M(n+1)^{G(1)} = M(n)$. Let $F: M(n+1) \to M(n+1,k)$ be the canonical map. Since F is equivariant, $F(M(n)) \subset M(n+1,k)^{G(1)} = M(n,k)$. Hence, F restricts to a map $F: M(n+1) - M(n) \to M(n+1,k) - M(n,k)$. Since F is stratified, it maps the normal bundle of $M(n)$ transversely to the normal bundle of $M(n,k)$. Notice that $M(n+1,k) - M(n,k) = (M(1,k) - \{0\}) \times M(n,k)$. Let $p\colon (M(1,k) - \{0\}) \times M(n,k) \to S^{dk-1}$ be projection onto the first factor followed by radial projection onto the unit sphere $S^{dk-1} \subset M(1,k) - \{0\}$. The map p is invariant with respect to the $G(n)$-action on $M(n+1,k) - M(n,k)$. Hence, $p \circ F\colon M(n+1) - M(n) \to S^{dk-1}$ is also invariant with respect to the $G(n)$-action. Let $x_0 \in S^{dk-1}$ be a regular value of $p \circ F$. Then $V = (p \circ F)^{-1}(x_0)$ is a $G(n)$-invariant submanifold of $M(n+1,k)$ with trivial normal bundle. Let $\bar{V}$ be the closure of V. It is easy to verify that the action on $\bar{V}$ is k-axial and that $\partial\bar{V} = M$. Furthermore, F maps $\bar{V}$ to $Rx_0 \times M(n,k)$, where Rx_0 is the line in $M(1,k)$ through x_0. Hence, $\bar{V}$ is a pullback of $M(n,k)$. If M is stably parallelizable, then by the previous proposition $T(B)$ is trivial and hence, $M(m)$ is stably parallelizable for all $m \geq k$. In particular, since $M(n+1)$ is stably parallelizable and the normal bundle of $\bar{V} \subset M(n+1)$ is trivial, it follows that $\bar{V}$ is stably parallelizable.

□

<u>Proof 2</u>: This next proof only works when the fixed point set is non-empty and M is closed and stably parallelizable. In this case we know by Proposition 2 in section V.C that there is a degree one stratified map $F\colon M \to S$ where S is a linear sphere of the same dimension as M. F can be covered by a map of the stable normal bundles since TM is stably trivial. Thus, F is a "transverse linear isovariant normal map" in the sense of Browder and Quinn [4]. According to them, the "isovariant normal invariant of F" is given by the homotopy class of a map $L \to G/O$ where $L = S/G(n)$. But L is

contractible. Hence, F is normally cobordant to the identity. Sewing in a linear disk along id: $S \to S$, the result follows. Even if the fixed point set is empty, this argument will (usually) work when M is a homology sphere. □

Proof 3: If $x \in H_+(k)$, let $e_1(x) \geq e_2(x) \geq \cdots \geq e_k(k) \geq 0$ be the eigenvalues of x, arranged in decreasing order. Let $T_i \subset H_+(k)$ be the manifold defined by

$$T_i = \{x | e_i(x) > e_{i+1}(x) = e_{i+2}(x) = \ldots = e_k(x)\}.$$

Thus,

$$\begin{aligned} \partial T_i &= \{x | e_i(x) > 0, \ e_{i+1}(x) = \ldots = e_k(x) = 0\} \\ &= H_+(k)_i . \end{aligned}$$

For example, T_0 is the ray of scalar matrices in $H_+(k)$.

Next, consider the map $s: H_+(k) \times [0,\infty) \to H_+(k)$ defined by $s(x,t) = x + tI$, where I is the identity matrix. The eigenvalues of $s(x,t)$ are the roots of $\det(x+(t-\lambda)I)$. It follows that $e_j(s(x,t)) = e_j(x) + t$. Suppose that x has rank i. Then $e_{i+1}(s(x,t)) = \ldots = e_k(s(x,t)) = t$, i.e., $s(x,t) \in T_i$. In other words, $s(H_+(k)_i \times [0,\infty)) = T_i$.

Now suppose that $M = f^*(M(n,k))$, where $f: B \to H_+(k)$. We can move f by a small stratified homotopy so that it is transverse to each T_i. Define $A \subset B \times H_+(k) \times [0,\infty)$ by $A = \{(b,x,t) | f(b) = s(x,t)\}$. In other words, A is the formal pullback of $s: H_+(k) \times [0,\infty) \to H_+(k)$ via f,

$$\begin{array}{ccc} A & \xrightarrow{\tilde{f}} & H_+(k) \times [0,\infty) \\ \downarrow & & \downarrow s \\ B & \xrightarrow{f} & H_+(k) \end{array}$$

Notice that $A \cap (B \times H_+(k) \times \{0\}) \cong B$ and that the restriction of $\tilde{f}$ to this intersection can be identified with f. Since f is transverse to each T_i, it follows that A has the structure of a local orbit space modeled on $H_+(k)$. For $i < k$, the i-stratum of A can be identified with $f^{-1}(T_i)$. Thus, the stratification on A is such that $\tilde{f}: A \to H_+(k) \times [0,\infty)$ is stratified.

Let $p_1: H_+(k) \times [0,\infty) \to H_+(k)$ be projection on the first factor and let $g = p_1 \circ f$. Set $V = g^*(M(n,k))$. The boundary of V is the part of V lying over $A \cap (B \times H_+(k) \times \{0\})$. Thus, $\partial V = M$.

Finally, consider the projection $q: A \to B$. If $b \in B$, $q^{-1}(b) = \{(x,t) | s(x,t) = f(b)\} \cong [0,\epsilon]$ where ϵ is the maximum t such that $f(b) - tI \in H_+(k)$. It follows that as a topological space A is homeomorphic to $B \times [0,1]$. Thus, if TB is trivial, then TA is also trivial. Hence, if M is stably parallelizable, then so is V. $\square$

Remark: I prefer the third proof since it gives the clearest picture of the orbit space A. Also, in this construction we have an extension of the classifying map $f: B \to H_+(k)$ to a classifying map $g: A \to H_+(k)$.

Using this and an argument similar to the proof of Proposition 12 in Chapter V.C, we have the following

Corollary 4: *Suppose that* $n \geq k$ *and that* $G(n)$ *acts* k-*axially on a homology sphere* $\Sigma^{dkn+m-1}$. *Then* Σ *equivariantly bounds a parallelizable* k-*axial* $G(n)$ *manifold* V *and there is a stratified map* $F: (V,\Sigma) \to (D,S)$, *where* D *is the unit disk in* $M(n,k) \times R^m$. *Moreover, except in the case where* $G(n) = O(n)$, $m = 0$, k *is even and* n *is odd*, F *can be chosen to be of degree one*.

C. The calculation of $\theta^d(k,n,m)$.

As usual suppose that $G(n)$ acts k-axially on a homotopy sphere Σ of dimension $dkn + m - 1$. Let V be the parallelizable manifold with $\partial V = \Sigma$, and with $F: (V,\Sigma) \to (D,S)$ a stratified normal map. Let $f: (A,B) \to (K,L)$ be the induced map of orbit spaces. To avoid the difficulties of 4-dimensional surgery we need to assume that no stratum of A has dimension 4. This will be implied, for example, by the hypotheses that

$$m \neq 0,4$$

and

$$m + d(k-1) + 1 > 4.$$

These conditions will be assumed for the remainder of this section.

Let us first discuss the easy case where $G(n) = U(n)$ or $Sp(n)$. Pick a framing $\psi: TA \to A \times R^N$. Consider the map of fixed point sets $f_0: (A_0,B_0) \to (K_0,L_0)$. Since f is stratified the normal bundle A_0 in A is the pullback of the normal bundle $N_0(K)$ of K_0 in K. Hence, ψ induces a framing $\psi_0: TA_0 \oplus f_0{}^*(N_0(K)) \to R^N$. Since $f_0|B_0$ is a homology isomorphism, (f_0,ψ_0) is a surgery problem in the sense of Wall [9]. Since $(K_0,L_0) = (D^m,S^{m-1})$, the surgery obstruction lies in

$$L_m(1) = \begin{cases} 1/8 \text{ the index of } f_0 & ; \quad m \equiv 0(4) \\ \text{Kervaire invariant of } f_0 ; & \quad m \equiv 2(f) \\ 0 & ; \quad \text{otherwise.} \end{cases}$$

Let $\sigma_0 \in L_m(1)$ be this obstruction. The main result of [5] is the following.

Theorem 5: Let $d = 2$ or 4. The map $\sigma_0: \theta^d(k,n,m) \to L_m(1)$ is a homomorphism. If k is even, σ_0 is an isomorphism. If k is odd, σ_0 is the zero map.

For further details in the unitary and symplectic cases see [5]. In the orthogonal case there is some difference between the cases where $k \neq 2$ and where $k = 2$ (the knot manifold case). The basic reason for this is that for $k \neq 2$, each stratum of the linear orbit space K has either trivial fundamental group or fundamental group $Z/2$; while for $k = 2$, $\pi_1(K_1) = Z$. So let us first discuss the special case of knot manifolds.

Suppose that a homotopy sphere Σ^{2n+m-1} supports a biaxial $0(n)$ action with orbit space B. First, consider the case where n is even. Then $B_0 \underset{Z}{\sim} S^{m-1}$ and $\partial B \underset{Z}{\sim} S^{m+1}$. Suppose that Σ equivariantly bounds a contractible manifold V with orbit space A. Then $A_0 \underset{Z}{\sim} D^m$, $A_1 \cup A_0 \underset{Z}{\sim} D^{m+2}$. It follows that B_0 is a "homology slice knot". More precisely, let $\tilde{C}_{m-1}$ denote the group of knotted homology $(m-1)$-spheres in homology $m+1$ spheres up to "homology knot cobordism", (where the homology $(m+1)$-sphere is required to bound a contractible manifold). It is fairly clear that for n even, $\theta^1(2,n,m) = \tilde{C}_{m-1}$. On the other hand, it is well-known that for $m-1 \neq 1,3$, $\tilde{C}_{m-1} = C_{m-1}$ where C_{m-1} is the ordinary knot cobordism group. Also, by work of Kervaire and Levine, $C_{m-1} = 0$ if $m-1$ is even and not equal to 2, and for $m-1$ odd and $m-1 \neq 1,3$, $C_{m-1} = G_\epsilon$, $\epsilon = (-1)^{m/2}$, where G_ϵ is the group of "cobordism classes of Seifert matrices" as defined by Levine in [8]. Thus, for n even, $m \neq 2,4$, $\theta^1(2,n,m) = G_\epsilon$.

For n odd, $B_0 \underset{Z/2}{\sim} S^{m-1}$, $\partial B \underset{Z}{\sim} S^{m+1}$ and $D_1(\Sigma) \underset{Z}{\sim} S^{m+1}$. Here $D_1(\Sigma)$

is the double branched cover of ∂B along B_0. If Σ equivariantly bounds a contractible manifold V, then its orbit space A satisfies similar conditions. It follows that for n odd, $\theta^1(2,n,m) = C_{m-1}(Z/2)$, where $C_{m-1}(Z/2)$ is defined as follows. Consider pairs $(\partial B, B_0)$ where

(1) B_0 is a $Z/2$-homology $(m-1)$-sphere.

(2) ∂B is an integral homology $(m+1)$-sphere which bounds a contractible manifold.

(3) The double branched cover of ∂B along B_0 is an integral homology sphere.

The equivalence relation is "Z/2-homology knot cobordism" such that the double branched cover of the cobordism is an "integral homology h-cobordism." The set of such equivalence classes is an abelian group denoted by $C_{m-1}(Z/2)$. Levine told me that $C_{m-1}(Z/2)$ is the "other knot cobordism", i.e., for $m-1$ even and $m-1 \neq 2$, $C_{m-1}(Z/2) = 0$; while for $m-1$ odd and $m-1 \neq 1,3$, $C_{m-1}(Z/2) = G_\epsilon$ where $\epsilon = (-1)^{m/2+1}$. The proof of this result will appear in [6]. (Note that $C_{m-1}(Z/2) \cong C_{m+1}$.) Therefore we have the following.

<u>Theorem 6</u>: <u>If</u> $m > 4$, <u>then</u>,

$$\theta^1(2,n,m) = \begin{cases} G_{+1}; & m+2n \equiv 0 \pmod 4 \\ 0 \ ; & m+2n \equiv 1 \pmod 4 \\ G_{-1}; & m+2n \equiv 2 \pmod 4 \\ 0 \ ; & m+2n \equiv 3 \pmod 4. \end{cases}$$

<u>Remark</u>: For n even, this result is essentially due to Bredon [2], [3].

It turns out that for $k \neq 2$, the concordance class of the $O(n)$-action on Σ is determined by the simply connected surgery obstructions of the map $(V^{O(n)},\Sigma^{O(n)}) \to (D^{O(n)},S^{O(n)})$ and of the map $(V^{O(n-1)},\Sigma^{O(n-1)}) \to (D^{O(n-1)},S^{O(n-1)})$, i.e., only the obstructions on the bottom two strata matter (all the others either vanish or are indeterminacy). If n is even the first map is an integral homology equivalence on the boundary and the second is a homology equivalence with coefficients in $Z_{(2)}$. If n is odd, the reverse is true. Notice that the dimension of $V^{O(n)}$ is m and that the dimension of $V^{O(n-1)}$ is $m + k$. If m is odd, there is no surgery obstruction for $(V^{O(n)},\Sigma^{O(n)}) \to (D^{O(n)},S^{O(n)})$. If $m \equiv 2 \pmod 4$, the only possible obstruction is the Kervaire invariant. If $m \equiv 0 \pmod 4$ and n is even, the obstruction is one eighth the index of the intersection form on $H_{1/2m}(V^{O(n)},\Sigma^{O(n)};Z)$. If $m \equiv 0 \pmod 4$ and n is odd, the intersection form is only nonsingular over $Z_{(2)}$. Hence, the obstruction lies in the Witt ring of even symmetric bilinear forms over $Z_{(2)}$. Similar remarks apply to the surgery obstruction for $(V^{O(n-1)},\Sigma^{O(n-1)}) \to (D^{O(n-1)}, S^{O(n-1)})$. There are now several different cases.

First of all, just as in the unitary and symplectic cases, for k odd, these obstructions always vanish (by "looking up" one or two strata). Hence, if k is odd, Σ is always concordant to the linear action.

If k is even and m is odd, then both $V^{O(n)}$ and $V^{O(n-1)}$ are odd dimensional; hence there are no surgery obstructions.

If $k \equiv 0 \pmod 4$, then $\dim V^{O(n-1)} = \dim V^{O(n)} + k = m + k \equiv m \pmod 4$. Hence, if $m \equiv 0 \pmod 4$ we have two even symmetric forms, one for $V^{0(n)}$ and one for $V^{0(n-1)}$, one of these will be nonsingular over Z while the other will be nonsingular over $Z_{(2)}$. If $m \equiv 2 \pmod 4$, then $\dim V^{0(n-1)} \equiv 2 \pmod 4$. Thus, in this case we are only concerned with (possibly two) Kervaire invariants. However, the Kervaire invariant of $V^{0(n-1)}$ automatically vanishes.

If $k \equiv 2 \pmod 4$, then $\dim V^{0(n-1)} = \dim V^{0(n)} + k = m + k \equiv m + 2 \pmod 4$. Hence, if m is even, exactly one of the manifolds $V^{0(n-1)}$ or $V^{0(n)}$ will have dimension divisible by four.

Let W be the Witt ring of nonsingular even, symmetric, bilinear forms over $Z_{(2)}$. The Witt ring of nonsingular, even, symmetric, bilinear forms over Z is isomorphic to the integers, where the isomorphism is defined by $a \to \frac{1}{8}$ (index of a); however, W is much larger. If $a \in W$, it can be represented by an integral matrix A of odd determinant. This determinant modulo 8 depends only on a. Consider the homomorphism $\psi\colon W \to Z/2$ defined by $\psi(a) = 0$ if $\det A = \pm 1 \pmod 8$ and $\psi(a) = 1$ if $\det(A) = \pm 3 \pmod 8$. $\psi(a)$ is the Kervaire invariant of a. Let $\bar{W} = \ker \psi$. Then $\theta^1(k,n,m)$ is calculated as follows.

<u>Theorem 7</u>:

(1) <u>If</u> k <u>is odd, then</u> $\theta^1(k,n,m) = 0$.

(2) <u>If</u> $k \equiv 0$ <u>(mod</u> 4), <u>then</u>

$$\theta^1(k,n,m) = \begin{cases} Z + \bar{W}\,; & m \equiv 0 \pmod 4 \\ 0\,; & m \equiv 1 \pmod 4 \\ Z/2\,; & m \equiv 2 \pmod 4 \\ 0\,; & m \equiv 3 \pmod 4\,. \end{cases}$$

(3) <u>If</u> $k \equiv 2$ (<u>mod</u> 4) <u>and</u> $k \neq 2$, <u>then</u>

$$\theta^1(k,n,m) = \begin{cases} Z \;; & m + 2n \equiv 0 \pmod 4 \\ 0 \;; & m + 2n \equiv 1 \pmod 4 \\ W \;; & m + 2n \equiv 2 \pmod 4 \\ 0 \;; & m + 2n \equiv 3 \pmod 4 \;. \end{cases}$$

The details of this calculation will appear in [6].

If $k \equiv 2 \pmod 4$, these calculations show that up to concordance every k-axial action on a homotopy sphere is obtained by restricting some biaxial action (that is, a knot manifold) to a subgroup. More precisely, if $k = 4\ell + 2$ and Σ admits a k-axial $O(n)$ action, then up to concordance the action is the restriction of a biaxial $O(m)$-action where $m = (2\ell+1)n$. (Notice that there are natural epimorphisms $G_{+1} \to Z$ and $G_{-1} \to W$.)

On the other hand for $k \equiv 0 \pmod 4$ and $m \equiv 0 \pmod 4$, the group $\theta^1(k,n,m)$ contains new examples of actions (that is, actions which are not restrictions of biaxial actions). At present our construction of these examples is not very explicit as it involves doing repeated surgery. It would be interesting to have nice descriptions of these actions. For example, the question arises: is there a nice description of a 4-axial $O(n)$-action so that the index of $V^{O(n)}$ is not equal to the index of $V^{O(n-1)}$?

As a further corollary to the calculations in Theorems 6 and 7 we have the following.

<u>Theorem 8</u>: <u>Let</u> $\alpha: \theta^d(k,n,m) \to \theta^d(k,n-1,m+dk)$ <u>be the homomorphism defined by restricting a</u> $G^d(n)$-<u>action to</u> $G^d(n-1)$. <u>Then</u> α <u>is an</u>

isomorphism.

This result is particularly interesting in the orthogonal case. It is also possible to compute many other homomorphisms induced by restriction of the action to a subgroup, but I shall not do it here.

References

[1] G. Bredon, Introduction to Compact Transformation Groups, Academic Press, New York, 1972.

[2] ________, Regular 0(n)-manifolds, suspension of knots, and knot periodicity, Bull. Amer. Math. Soc. 79 (1973), 87-91

[3] ________, Biaxial actions, notes, Rutgers University, 1973.

[4] W. Browder and F. Quinn, A surgery theory for G-manifolds and stratified sets, Manifolds, University of Tokyo Press, 1973, 27-36.

[5] M. Davis and W. C. Hsiang, Concordance classes of regular U(n) and Sp(n) actions on homotopy spheres, Ann. of Math., 105 (1977), 325-341.

[6] M. Davis, W. C. Hsiang and J. Morgan, Concordance classes of regular 0(n)-actions on homotopy spheres, (to appear).

[7] M. Kervaire and J. Milnor, Groups of homotopy spheres I, Ann. of Math., 77 (1963), 504-537.

[8] J. Levine, Knot cobordism groups in codimension two, Comment. Math. Helv., 44 (1969), 229-244.

[9] C. T. C. Wall, Surgery on Compact Manifolds, Academic Press, New York, 1970.

APPENDIX 1: BIAXIAL ACTIONS OF SO(3), G_2, AND SU(3) ON HOMOTOPY SPHERES

<u>Preliminaries</u>: We begin by recalling some well-known facts. Let $\mathbb{H}$ be the quaternions and let C be the Cayley numbers. Let S^3 denote the quaternions of length one ($S^3 = Sp(1)$). There is an epimorphism $\varphi: S^3 \to \mathrm{Aut}(\mathbb{H})$ defined by $\varphi(x)y = xyx^{-1}$. The kernel of φ is $\{\pm 1\}$. Thus,

$$\mathrm{Aut}(\mathbb{H}) \cong SO(3).$$

The action of SO(3) on $\mathbb{H}$ leaves the real numbers fixed and it acts on the perpendicular 3-dimensional subspace by rotation. In other words, the action of SO(3) on $\mathbb{H}$ is equivalent to $1\rho_3 + 1$.

The group of automorphisms of the Cayley numbers is called G_2. It is a compact simply connected simple Lie group of dimension 14. As before, every automorphism fixes the $\mathbb{R}1 \subset C$ and G_2 acts orthogonally and irreducibly on the perpendicular 7-dimensional subspace. This action of G_2 on $\mathbb{R}^7$ is called the <u>standard representation of</u> G_2.

As a module over the complex numbers $C \cong \mathbb{C}^4$. The subgroup of G_2 which fixes $\mathbb{C}1$ is isomorphic to SU(3) and $N(SU(3))/SU(3) \cong O(1)$. The induced action of O(1) on $\mathbb{C}1$ is by complex conjugation. As a module over the quaternions $C \cong \mathbb{H}^2$. The subgroup which fixes $\mathbb{H}1$ is isomorphic to SU(2) and $N(SU(2))/SU(2) \cong SO(3)$. Of course, the induced action of SO(3) on $\mathbb{H}1 = C^{SU(2)}$ is by $\mathbb{H}$-automorphisms. Also note that the restriction of G_2 on C to SU(3) is equivalent to $1\rho_3 + 2$; while the restriction to SU(2) is equivalent to $1\rho_2 + 4$. Here ρ_i denotes the standard representation of SU(i) on $\mathbb{C}^i$.

Now, let $G' = O(3)$, $SO(7)$ or $U(3)$ as $G = SO(3)$, G_2 or $SU(3)$. Let ρ': $G' \times V \to V$ denote the standard representation of G' on R^3, R^7, or $\mathbb{C}^3$ and let ρ: $G \times V \to V$ be the restriction of ρ' to G. The orbits of G on V are equal to the orbits of G' on V (in both cases the nonsingular orbits are spheres of codimension one). Thus, $V/G = V/G' \cong [0,\infty)$. Also, it is trivial to check that the actions 2ρ: $G \times V^2 \to V^2$ and $2\rho'$: $G' \times V^2 \to V^2$ have the same orbits. Thus, $V^2/G = V^2/G'$. From this we immediately deduce the following.

<u>Proposition 1</u>: <u>Let</u> $G' = O(3)$, $SO(7)$ <u>or</u> $U(3)$ <u>as</u> $G = SO(3)$, G_2 <u>or</u> $SU(3)$. <u>Then every smooth biaxial</u> G'-<u>action restricts to a biaxial</u> G-<u>action with the same orbit space</u>.

However, it is false that every biaxial G-action extends to a biaxial G'-action. Furthermore, if such an extension exists it may not be unique. Recall that it was shown in Chapter III that for $G' = O(3)$ or $SO(7)$, a biaxial G'-action on a homotopy sphere is determined by its orbit space. For $G = SO(3)$ or G_2, we will give examples in the next section of an infinite number of biaxial G actions on homotopy spheres (of dimension 7 or 15) with the same orbit space as the linear action.

<u>Some Concrete Examples</u>: It is well-known that $\pi_3(SO(4)) \cong \mathbb{Z} + \mathbb{Z}$ and that $\pi_7(SO(8)) \cong \mathbb{Z} + \mathbb{Z}$. Explicit isomorphisms are given as follows. Define φ_{ij}: $S^3 \to SO(4)$ by $\varphi_{ij}(x)y = x^i y x^j$ where $y \in \mathbb{H} \cong R^4$. Similarly, define $\hat{\varphi}_{ij}$: $S^7 \to SO(8)$ by $\hat{\varphi}_{ij}(x)\cdot y = (x^i y)x^j$. Here $x \in S^7$ is regarded as a unit Cayley number and $y \in C \cong R^8$. Then $(i,j) \to [\varphi_{ij}]$ and $(i,j) \to [\hat{\varphi}_{ij}]$ define isomorphisms $\mathbb{Z} + \mathbb{Z} \cong \pi_3(SO(4))$ and

$\mathbb{Z} + \mathbb{Z} \cong \pi_7(SO(8))$. Let ξ_{ij} be the 4-plane bundle over S^4 with characteristic map φ_{ij}. Similarly, let $\hat{\xi}_{ij}$ be the 8-plane bundle over S^8 with characteristic map $\hat{\varphi}_{ij}$. Let $E(\xi_{ij})$ and $E(\hat{\xi}_{ij})$ be the total spaces of the associated disc bundles and let $E_0(\xi_{ij})$ and $E_0(\hat{\xi}_{ij})$ be the total spaces of the associated sphere bundles. Let $D^4 \subset \mathbb{H}$ be the unit disk. Then $E(\xi_{ij})$ is the union of two copies of $D^4 \times D^4$ via the attaching map $\theta_{ij}\colon S^3 \times D^4 \to S^3 \times D^4$ defined by $\theta_{ij}(x,y) = (x,\varphi_{ij}(x)y)$. If $\alpha \in SO(3) = \mathrm{Aut}(\mathbb{H})$, then

$$\begin{aligned}\theta_{ij}(\alpha x,\alpha y) &= (\alpha x,\varphi_{ij}(\alpha x)\,\alpha y)\\ &= (\alpha x,(\alpha x)^i(\alpha y)(\alpha x)^j)\\ &= (\alpha x,\alpha(x^i y x^j))\\ &= (\alpha x,\alpha(\varphi_{ij}(x)y)).\end{aligned}$$

Consequently, the attaching map θ_{ij} is equivariant with respect to the canonical biaxial SO(3) action on $S^3 \times D^4$. Therefore, SO(3) acts biaxially on ξ_{ij} through bundle maps. In a similar fashion the attaching map $\hat{\theta}_{ij}\colon S^7 \times D^8 \to S^7 \times D^8$ defined by $(x,y) \to (x,\hat{\varphi}_{ij}(x)y)$ is equivariant with respect to the canonical biaxial G_2 action. Therefore we have proved the following.

Theorem 2: Every 4-plane bundle over S^4 canonically has the structure of an SO(3)-vector bundle. The action on the base is the linear action of SO(3) on $S^4 = \mathbb{H}P^1$. Also, every 8-plane bundle over S^8 canonically has the structure of a G_2-vector bundle. On the base the action is the linear action of G_2 on $S^8 = CP^1$.

Of course, the manifolds $E_0(\xi_{ij})$ were first considered by Milnor

[6]. The Euler class of the bundle is (i+j) times a generator of $H^4(S^4;\mathbb{Z})$. Thus, if $i + j = \pm 1$, $E_0(\xi_{ij})$ is homeomorphic to S^7. (Notice the $\xi_{1,0}$ is the Hopf-bundle.) Milnor showed that for some pairs (i,j) with $i + j = 1$, $E_0(\xi_{ij})$ was not diffeomorphic to S^7. His calculations were later refined by Eells and Kuiper [2]. In fact, they showed that $E_0(\xi_{2,-1})$ was the generator of θ_7 the group of exotic 7-spheres ($\theta_7 \cong \mathbb{Z}/28$). (As we pointed out in section F of Chapter III the SO(3) action of $E_0(\xi_{2,-1})$ was first noticed in [3].) In a similar fashion, if $i + j = 1$, the manifolds $E_0(\hat{\xi}_{ij})$ are homotopy 15-spheres. It is known [5] that $\theta_{15} = bP_{16} + \mathbb{Z}/2$, where bP_{16} is the subgroup of homotopy spheres which bound π-manifolds ($bP_{16} \cong \mathbb{Z}/8128$). It follows from work of Wall [8], that if a homotopy 15-sphere bounds a 7-connected 16-manifold, then it also bounds a π-manifold. Also, it follows from [2], that $E_0(\hat{\xi}_{2,-1})$ is the generator of bP_{16}.

Now we consider the actions on these bundles in more detail. Since $(D^4 \times D^4)^{SO(3)} = \{(x,y) \mid x \text{ and } y \text{ are real}\}$ and since $(D^4 \times D^4)^{SO(2)} = \{(x,y) \mid x \text{ and } y \text{ are complex}\}$, we see that $E(\xi_{ij})^{SO(3)}$ is a band with (i+j) half twists (i.e., a cylinder if $i + j$ is even and a Möebius band if $i + j$ is odd); while $E(\xi_{ij})^{SO(3)}$ is the 2-disk bundle over S^2 with Euler class = (i+j) times a generator of $H^2(S^2;\mathbb{Z})$. In a similar fashion, for the G_2-action on $E(\hat{\xi}_{ij})$ we have that

$$E(\hat{\xi}_{ij})^{SU(2)} = E(\xi_{ij})$$

$$E(\hat{\xi}_{ij})^{SU(3)} = E(\xi_{ij})^{SO(2)}$$

$$E(\hat{\xi}_{ij})^{G_2} = E(\xi_{ij})^{SO(3)}.$$

These equations lead us to suspect that the G_2-action on $E(\hat{\xi}_{ij})$ is determined by the SO(3)-action on $E(\xi_{ij})$. More generally, one has the following.

Theorem 3: The functor $M \to M^{SU(2)}$ is an equivalence of categories from the category of biaxial G_2-actions and stratified maps to the category of biaxial SO(3)-actions and stratified maps. Moreover, this functor takes integral homology spheres to integral homology spheres.

It is trivial to check that $M/G_2 \cong M^{SU(2)}/SO(3)$. It then follows from the Covering Homotopy Theorem and the fact that the normal orbit bundles of M have the same structure groups as the corresponding normal orbit bundles of $M^{SU(2)}$ that the natural inclusion $\mathrm{Hom}(M,N) \to \mathrm{Hom}(M^{SU(2)},N^{SU(2)})$ is a homeomorphism. Here Hom(,) denotes the space of (equivariant) stratified maps. There are many similar theorems of this type.*

It is not difficult to see that the orbit space of SO(3) on $E(\xi_{ij})$ is homeomorphic to a 5-disk. The boundary of this 5-disk is $D^4_+ \cup D^4_-$ where D^4_- is the union of the singular orbits and where D^4_+ is the orbit space of the sphere bundle $E_0(\xi_{ij})$. The fixed set is a band with $i + j$ half twists embedded in D^4_-. This band can be pushed into S^3 to be a (possibly non-orientable) Siefert surface for the fixed set of $E_0(\xi_{ij})$. The boundary of a twisted band in S^3 is a torus link of type $(2,i+j)$. In particular, if $i + j = \pm 1$, this is an

---

*Let ψ: $G \times R^m \to R^m$ be a representation with principal orbit type G/H. Let $K = N(H)/H$ and let φ: $K \times (R^m)^H \to (R^m)^H$ be the induced representation. Then it follows from Schwarz's proof of the Covering Homotopy Theorem that for "most" ψ the functor $M \to M^H$ is an equivalence from the category of G-actions modeled on ψ to the category of K-actions modeled on φ.

unknotted circle. We observe that more generally the orbit space of SO(3) on $E_0(\xi_{ij})$ is the isomorphic to the orbit space of SO(3) on the Brieskorn manifold $\Sigma^7(i+j,2,2,2,2)$. Similar remarks apply to G_2 on $E_0(\hat{\xi}_{ij})$. Thus, we have the following theorem.

Theorem 4: Suppose $i + j = 1$. Then $E_0(\xi_{ij})$ (respectively, $E_0(\hat{\xi}_{ij})$) is a homotopy 7-sphere (respectively, 15-sphere). The orbit space of SO(3) on $E_0(\xi_{ij})$ (respectively, G_2 on $E_0(\hat{\xi}_{ij})$) is isomorphic to the orbit space of the linear biaxial SO(3) action on S^7 (respectively, G_2 on S^{15}), that is to say, the orbit space is a 4-disk with fixed point set an unknotted circle embedded in the boundary.

We can also obtain biaxial SO(3) or G_2 actions over the linear orbit space by taking equivariant equivariant connected sums (at fixed points) of the above examples.

Let us simplify the notation. Let

$$\Sigma^7_h = E_0(\xi_{h,h-1})$$

and

$$\Sigma^{15}_h = E_0(\hat{\xi}_{h,h-1}).$$

Then Σ^7_1 and Σ^{15}_1 are equivalent to the linear actions on S^7 and S^{15}. Since the disk bundles are spin manifolds, one can define the μ invariant of their boundaries. It is computed by Eells and Kuiper [2] as

$$\mu(\Sigma^7_h) = \tfrac{1}{2}h(h-1)/28$$

$$\mu(\Sigma^{15}_h) = \tfrac{1}{2}h(h-1)/8128 \quad .$$

If a closed spin manifold admits an S^1-action its $\hat{A}$-genus vanishes. It follows that these rational numbers are invariants of the SO(3) (or G_2) action on Σ_h^7 (or Σ_h^{15}). (See page 351 in [1].) In particular the numerator $\frac{1}{2}h(h-1)$ is an invariant of the action. Notice that for $h = 2$, $\frac{1}{2}h(h-1) = 1$. As G = SO(3) or G_2, let d = 3 or 7. Let G_{2d+1} be the cyclic group of biaxial G-actions generated by Σ_2^{2d+1} (under connected sum). Since the μ-invariant is additive, it follows from the above remarks that G_{2d+1} is infinite cyclic. We state this as the following.

Theorem 5: As G = SO(3) or G_2, let d = 3 or 7. There is an infinite cyclic group of oriented biaxial G-actions on homotopy (2d+1)-spheres so that the orbit space is isomorphic to the orbit space of the linear action on S^{2d+1}.

In fact G_{2d+1} is the group of all biaxial G-actions over the linear orbit space. We shall indicate the proof of this in the next section. Next consider what happens when we restrict these SO(3)-actions to SO(2) and these G_2-actions to SU(3). The restriction of a biaxial SO(3)-action to SO(2) is biaxial, i.e., it is a semifree S^1-action with fixed point set of codimension 4. Clearly, $(\Sigma_h^7)^{SO(3)} = S^3$. Also, it is trivial to check that $\Sigma_h^7/SO(3) \cong S^6$. Recall that Haefliger defined a group $\Sigma^{6,3}$ of knotted 3-spheres in S^6 and proved in [4] that

$$\Sigma^{6,3} \cong \mathbb{Z}.$$

Also, Montgomery and Yang [7] proved that for every $\alpha \in \Sigma^{6,3}$ there is a unique semifree S^1-action on a homotopy seven sphere Σ so that

$\alpha = (\Sigma/S^1, \Sigma^{S^1})$. It is possible to compute that the SO(2) action on Σ_2^7 corresponds to a generator of $\Sigma^{6,3}$. Therefore, we have the following.

Theorem 6: *The homomorphism* $C_7 \to \Sigma^{6,3}$ *which sends the generator* Σ_2^7 *to* $((\Sigma_2^7)/S^1, (\Sigma_2^7)^{S^1})$ *is an isomorphism. Thus, each of Montgomery and Yang's semifree* S^1*-action on homotopy* 7*-spheres extends uniquely* to a biaxial* SO(3) *action*.

The orbit space of SU(3) on Σ_h^{15} is closely related to the orbit space of SO(2) on Σ_h^7. In fact, $\Sigma_h^{15}/SU(3)$ is a 7-disk and the singular set is isomorphic to $\Sigma_h^7/SO(2)$. Therefore, as a corollary to Theorem 6 we have the following theorem.

Theorem 7: *Given any* $\alpha \in \Sigma^{6,3}$ *there is a biaxial* SU(3) *action on a homotopy* 15*-sphere such that the singular set of the orbit space corresponds to* α.

Corollary 8: *There are biaxial* SU(3)*-actions on homotopy* 15*-spheres which do not extend to biaxial* U(3)*-actions*.

Proof: The orbit knot of a biaxial U(3) action has an orientable Siefert surface (which must be parallelizable). According to Haefliger the knotted 3-spheres which admit orientable Siefert surfaces form a subgroup of $\Sigma^{6,3}$ of index 2. □

Of course, Corollary 8 makes good sense since for G = SO(3) or G_2 the actions on m Σ_2^{2d+1} do not extend to biaxial G' actions

---

*Uniqueness will follow from results of next section.

($G' = O(3)$ or $SO(7)$). Since Σ_2^{15} is a generator of bP_{16}, the forgetful homomorphism $G_{15} \to bP_{16}$ is onto. At this point it still seems possible (although unlikely) that there are biaxial $SU(3)$-actions on homotopy 15-spheres which up to concordance are neither the restriction of a $U(3)$-action or of a G_2-action. Thus, for example, the question of the existence of a Σ^{15} with $\Sigma^{15} \notin bP_{16}$, such that Σ^{15} admits a biaxial $SU(3)$-action, remains open.

Structure Theorems and Classification Theorems:

Let $G = SO(3)$, G_2 or $SU(3)$. Since biaxial G-actions with trivial bundle of principal orbits do not necessarily pull back from their linear models, the question arises: is there some other universal G-manifold which classified them and if so what is it? The answer is somewhat surprising. Since $SO(3)$ is the group of automorphisms of the quaternions, it acts on the quaternionic projective plane. The action is clearly biaxial. In a similar fashion G_2 acts biaxially on the Cayley projective plane CP^2. Hence $SU(3)$ also acts (by restriction) on CP^2. It turns out that these are the universal examples we seek.

Let us consider these actions in more detail. First, notice that

$$(CP^2)^{SU(2)} = \mathbb{H}P^2$$

$$(CP^2)^{SU(3)} = (\mathbb{H}P^2)^{SO(2)} = \mathbb{C}P^2$$

$$(CP^2)^{G_2} = (\mathbb{H}P^2)^{SO(3)} = \mathbb{R}P^2.$$

Let X denote the orbit space of $SO(3)$ on $\mathbb{H}P^2$. By Theorem 3,

$$X \cong CP^2/G_2.$$

Also, let Y be the orbit space of $SU(3)$ on CP^2.

Proposition 9: As a stratified space X is isomorphic to the orbit space of the linear triaxial $SO(4)$ action on S^{11}, and Y is isomorphic to the orbit space of the linear triaxial $U(4)$ action on S^{23}. More explicitly,

1) $X_0 = \mathbb{R}P^2$, $X_0 \cup X_1 \cong S^4$, $X_2 = D^5$ and X_1 is a 2-plane bundle over $\mathbb{R}P^2$.

2) $Y_0 = \mathbb{C}P^2$, $Y_0 \cup Y_1 \cong S^7$, $Y_2 = D^8$ and Y_1 is a 3-plane bundle over $\mathbb{C}P^2$.

Proof: Consider $3\rho_4: SO(4) \times S^{11} \to S^{11}$. We have the inclusion $1\rho_1: Sp(1) \hookrightarrow SO(4)$ of a normal subgroup. Clearly, $S^{11}/Sp(1) = \mathbb{H}P^2$, and the induced action of $SO(4)/Sp(1) \cong SO(3)$ on the quotient is equivalent to the action described above. Consequently, $X \cong S^{11}/SO(4)$. The result for Y is a little more complicated. We shall prove that Y_1 and Y_0 are correct and leave it to the reader to show that Y_2 is a disk. We have

$$
\begin{aligned}
\partial Y &= Y_1 \cup Y_0 \\
&= (CP^2)^{SU(2)}/U(1) \\
&= \mathbb{H}P^2/SO(2) \\
&\cong S^{11}/U(2)
\end{aligned}
$$

where $U(2)$ acts triaxially on S^{11}. This is the desired result. □

Theorem 10 (A Structure Theorem): Let $G = SO(3)$, G_2 or $SU(3)$. Let M be a biaxial G-manifold and suppose that the orbit bundle over the top stratum is a trivial fiber bundle. If $G = SO(3)$, then there is *a stratified map* $M \to \mathbb{H}P^2$. *If* $G = G_2$ *or* $SU(3)$ *then there is a stratified map* $M \to CP^2$.

Remark: There are stronger versions of this theorem analogous to the results in Chapter IV, Section A.3.

This theorem is proved in the same manner as the Structure Theorem of Chapter IV, that is, it is proved using twist invariants. The difference is the structure groups of the normal orbit bundles are not the same as they were for biaxial $O(3)$, $SO(7)$ or $U(3)$ actions. Let S_i denote the structure group of the i-normal orbit bundle. For example, for $G = SO(3)$, S_i is the group of orthogonal G-bundle automorphisms of

$$SO(3) \times_{SO(3-i)} M(3-i, 2-i).$$

We compare these structure groups for biaxial G-actions and the corresponding G'-actions in the following tables

	$G = SO(3)$	$G' = O(3)$
S_0	$O(2)$	$O(2)$
S_1	$O(2)$	$O(1) \times O(1)$
S_2	$SO(3)$	$O(2)$

	$G = SU(3)$	$G' = U(3)$
S_0	$U(2)$	$U(2)$
S_1	$U(2)$	$U(1) \times U(1)$
S_2	$SU(3)$	$U(2)$

The groups for $G = G_2$ and $G' = SO(7)$ are the same as in the first table.

The proof of Theorem 10 now follows from the next proposition as in Chapter III.

Proposition 11: *As $G = SO(3)$, G_2 or $SU(3)$ let N denote the "universal" G-manifold $\mathbb{H}P^2$, CP^2 or CP^2. Then for $i = 0,1,2$, the i-twist invariant can be defined for a biaxial G-manifold and the i-twist invariant for N is an S_i-equivariant homotopy equivalence to S_2.*

The proof of this proposition is left to the reader. The Structure Theorem now follows as in Chapter IV.

We now turn to classification. Let B be a local orbit space modeled on $H_+(2)$ and let $C_G(B)$ be the set of oriented equivalence classes of biaxial G-manifolds over B. Let $\hat{C}_G(B)$ be the subset consisting of those M with trivial bundle of principal orbits. Then as a corollary to Theorem 10, we have the following.

Theorem 12 (Classification Theorem): *There are bijections*

$$\hat{C}_{SO(3)}(B) \longleftrightarrow \hat{C}_{G_2}(B) \longleftrightarrow \pi_0(\mathrm{Hom}(B;X))/[B;SO(3)]$$

$$\hat{C}_{SU(3)}(B) \longleftrightarrow \pi_0(\mathrm{Hom}(B;Y))/[B;SU(3)].$$

Here, as usual, Hom (;) denotes the space of stratified maps.

If B is contractible (as is the case when the total space is a homotopy sphere), then there is no ambiguity introduced by choice of trivialization of the bundle of principal orbits and we can forget the terms $[B;SO(3)]$ and $[B;SU(3)]$.

Of course, the above theorem is not easy to work with, since it is generally very tricky and difficult to compute the set of homotopy classes of stratified maps. However, one can use it to prove the following result which we mentioned in the last section.

Theorem 13: Let L denote the linear orbit space $S^7/SO(3)$. Then $C_{SO(3)}(L) = G_7 \cong \mathbb{Z}$ and $C_{G_2}(L) = G_{15} \cong \mathbb{Z}$. Here G_{2d+1} is the group generated by Σ_2^{2d+1}.

Sketch of Proof: Let $f: L \to X$ be a stratified map. Then $L_0 = S^1$ and $X_0 = \mathbb{R}P^2$. The map $f|_{L_0} : L_0 \to X_0$ is null-homotopic since the normal bundle of L_0 in L is orientable (it is, in fact, trivial). Hence, after a homotopy we may assume $f|_{L_0}$ is constant. Next $\partial_0 L_1 = S^1 \times S^1$ and $\partial_0 X_1$ is a S^1-bundle over $\mathbb{R}P^2$. $\partial_0 L_1 \to \partial_0 X_1$ is a bundle map. The map composition $\partial_0 L_1 \to \partial_0 X_1 \subset X_1 \sim \mathbb{R}P^2$. is homotopic to the map $\beta(x,y) = \alpha(x)$ where $(x,y) \in S^1 \times S^1$ and $\alpha: S^1 \to \mathbb{R}P^2$ represents the generator of $\pi_1(\mathbb{R}P^2)$. Hence, we may assume $f|_{\partial_0 L_1}$ is β. Since $L_1 = D^2 \times S^1$, the problem now comes down to computing homotopy classes of maps $g: (D^2 \times S^1, S^1 \times S^1) \to \mathbb{R}P^2$ relative to $g|_{S^1 \times S^1} = \beta$. A little argument shows this set is $\pi_2(\Omega \mathbb{R}P^2) = \pi_3 \mathbb{R}P^2 \cong \mathbb{Z}$. (Here $g = f|_{L_1}$.) If the g represents 0, then f is homotopic to the map for the linear action. Although there were some choices, the above argument at least shows that $C_{SO(3)}(L)$ is a quotient of $\mathbb{Z}$. On the other hand, $\mathbb{Z} \cong G_7 \subset C_{SO(3)}(L)$. Therefore, $C_{SO(3)}(L) = C_{G_2}(L) \cong \mathbb{Z}$. Now a (complicated calculation shows that the map classifying Σ_2^7 represents a generator of $\pi_2(\Omega \mathbb{R}P^2)$.

Corollary 14: As $G = SO(3)$ or G_2 let $G' = O(3)$ or $SO(7)$ and let d = 3 or 7. Then every oriented biaxial G-action on a homotopy (2d+1)-sphere can be written uniquely as the connected sum of the restriction of a biaxial G'-action and some multiple of Σ_2^{2d+1} (the generator of Θ_{2d+1}).

Theorem 12 should have applications to other biaxial actions of SO(3), G_2 or SU(3) on homotopy spheres of other dimensions. For example, it appears that elements of $\pi_m(\Omega RP^2)$ give rise to biaxial SO(3)-actions on homotopy (6+m-1) spheres over the linear orbit space $S^{6+m-1}/SO(3)$. However, at this point, I am still unsure about the calculations.

References

[1] G. Bredon, Introduction to Compact Transformation Groups, Academic Press, New York, 1972.

[2] J. Eells and N.H. Kuiper, An invariant for certain smooth manifolds, Annali Mat. 60 (1962), 93-110.

[3] D. Gromoll and W. Meyer, An exotic sphere with nonnegative sectional curvature, Ann. of Math. 100 (1974) 447-490.

[4] A. Haefliger, Differentiable embeddings of S^n in S^{n+q} for $q > 2$. Ann. of Math., 83 (1966), 402-436.

[5] M. Kervaire and J. Milnor, Groups of homotopy spheres I, Ann. of Math., 77 (1963), 504-537.

[6] J. Milnor, On manifolds homeomorphic to the 7-sphere, Ann. of Math., 64 (1956), 399-405.

[7] D. Montgomery and C.T. Yang, Differentiable actions on homotopy seven spheres. Trans. Amer. Math. Soc. 122 (1966), 33-46.

[8] C.T.C. Wall, Classification of (n-1)-connected 2n-manifolds, Ann. of Math., 75 (1962), 163-189.

APPENDIX 2: THE HOMOLOGY OF THE STRATA IN THE O(n) CASE

Our goal is to prove the following result, which was stated as Lemma 6 in Chapter V.

Main Lemma: Let $F: M \to M'$ be a stratified map of k-axial O(n)-manifolds. Then a necessary and sufficient condition for F to be acyclic* (over Z) is that for each i with $0 \leq i \leq k+1$ and $n-i$ even, the induced map of double branched covers, $D_i(F): D_i(M) \to D_i(M')$ is acyclic.

Remark 1: We adopt the usual convention for the homology of the empty set, that is,

$$H_i(\phi;\Lambda) = \begin{cases} \lambda & ; \quad \text{for } i = -1 \\ 0 & ; \quad \text{for } i \neq -1 . \end{cases}$$

Notice that in the above lemma there is no hypothesis that any of the strata be nonempty. In particular, it is not assumed that $n \geq k$. Of course, as a consequence of the Main Lemma and the above convention one can deduce that if $F: M \to M'$ is acyclic, then a stratum of M is nonempty if and only if the corresponding stratum of M' is nonempty.

Remark 2: That condition that $D_i(F): D_i(M) \to D_i(M')$ be acyclic over Z implies that both of the maps $f|_{B_i}: B_i \to B_i'$ and $f|_{B_{i-1}}: B_{i-1} \to B_{i-1}'$ are acyclic over Z/2. Here, as usual, $f: B \to B'$ denotes the map of orbit spaces induced by F. That is to say, the Main Lemma implies

*Recall that a map is acyclic over a ring Λ if it induces an isomorphism on homology with coefficients in Λ.

Lemma 5 (the weak version of the Main Lemma) in Chapter V. (To see this consider the action of $Z/2 = O(i)/SO(i)$ on $D_i(F): D_i(M) \to D_i(M')$ and apply the Smith-Gysin sequence.) However, we need to use Lemma 5 in our proof; thus, at several points we need to use the fact that $H_*(f|_{B_i};Z)$ is an odd torsion group.

<u>Notation</u>: If $0 \leq j \leq k$, then

$M(j)$ is the complement in M of the union of open tubular neighborhoods of the strata of index less than j.

N_j is a tubular neighborhood of M_j in $M(j)$.

X_j is a regular neighborhood of $M_j \cup M_{j-1}$ in M, i.e., $X_j = N_j \cup N_{j-1}$.

Y_j is the intersection of N_j and N_{j-1}, i.e., $Y_j = N_j|_{\partial_{j-1}M_j}$.

P_j is ΣN_j, the unit sphere bundle associated to N_j.

Q_j is $\Sigma N_{j-1} - Y_j$, and

∇X_j is that portion of the boundary of X_j which is made up of the two sphere bundles, i.e., $\nabla X_j = P_j \cup Q_j$.

Picture of X_j

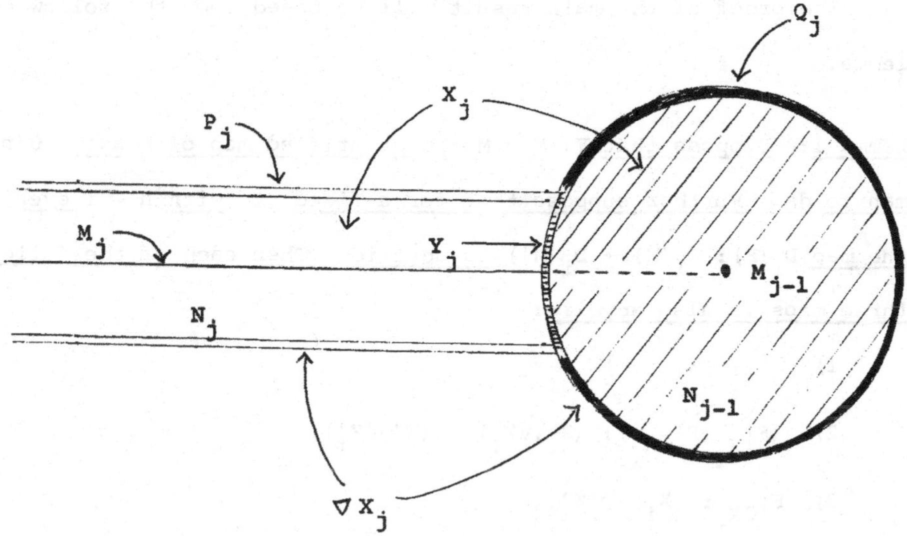

Legion

The manifolds N_j', X_j', Y_j', P_j', Q_j' and $\nabla X_j'$ are defined in a similar fashion using M'.

After an equivariant isotopy we may assume that the stratified map $F\colon M \to M'$ is a bundle map on the prescribed tubular neighborhoods.

Hence, we may assume that F maps each of the above submanifolds of M to corresponding submanifolds of M' and that F restricts to bundle maps $N_j \to N'_j$, $Y_j \to Y'_j$, $P_j \to P'_j$, $Q_j \to Q'_j$.

The proof of the main result will be based upon the following lemma.

Lemma 1: Suppose that $F: M \to M'$ is stratified map of k-axial $O(n)$-manifolds. Further suppose that for a fixed i with $n - i$ even, the map $D_i(F): D_i(M) \to D_i(M')$ is acyclic. Then each of the following three maps is also acyclic:

1) $F|_{X_i}$ $X_i \to X'_i$

2) $(F|_{X_i}, F|_{\nabla X_i}): (X_i, \nabla X_i) \to (X'_i, \nabla X'_i)$

3) $F|_{\nabla X_i}: \nabla X_i \to \nabla X'_i$.

First let us see how this lemma implies the Main Lemma. Rather than take M apart (or assemble it) one stratum at a time as in the $U(n)$ and $Sp(n)$ cases (Lemma 1 in Chapter V), the idea is to use the above lemma to take M apart (or assemble it) two strata at a time.

Proof of the Main Lemma: First we prove sufficiency. So suppose that $D_i(F): D_i(M) \to D_i(M')$ is acyclic whenever $n - i$ is even. For any integer j with $n - j$ even, set $J_j = M(j-1)$, i.e.,

$$J_j = M - \cup X_\ell$$

where the union is taken over all ℓ, with $\ell < j$ and $n - \ell$ even. Clearly, if n is even, then $J_0 = M$ and if n is odd $J_1 = M$. Let $t' = \min(n,k)$ be the index of the top nonempty stratum of M. Set $t = t'$ if $n - t'$ is even and $t = t' + 1$ if $n - t'$ is odd. Clearly,

$$J_t = X_t$$

(since J_t has at most two nonempty strata). Hence, by Lemma 1, $F|_{J_t}$ is acyclic. The proof now proceeds by downward induction. So, assume inductively that $F|_{J_{i+2}} : J_{i+2} \to J'_{i+2}$ is acyclic, where n - i is even. Consider the pair (J_i, X_{i+2}). By excision, $H_*(J_i, X_{i+2}) \cong H_*(J_{i+2}, \nabla X_{i+2})$. By Lemma 1, $F|_{X_{i+2}}$ and $F|_{\nabla X_{i+2}}$ are acyclic. Hence,

$$0 = H_*(F|_{J_{i+2}}, F|_{\nabla X_{i+2}})$$

$$\cong H_*(F|_{J_i}, F|_{X_{i+2}})$$

$$\cong H_*(F|_{J_i}).$$

This completes the inductive step and therefore proves sufficiency.

Next we prove necessity. So, suppose $F: M \to M'$ is acyclic, and that n - i is even. By Lemma 4 in Chaater V, tthe induced map of fixed point sets $F^{O(n-i)}: M^{O(n-i)} \to (M')^{O(n-i)}$ is also acyclic. Notice that this map is a stratified map of k-axial O(i)-manifolds. Recall that by definition $D_i(M)$ is the quotient (by SO(i)) of the union of the top two strata of $M^{O(n-i)}$. In particular, $D_i(M) = D_i(M^{O(n-i)})$ and $D_i(F) = D_i(F^{O(n-i)})$. Thus, to prove that $D_i(F)$ is acyclic it suffices to consider the case where $F: M \to M'$ is an acyclic stratified map of k-axial O(i)-manifolds, with $i < k$. The proof now proceeds by induction. If i = 0 or 1, then $M = D_i(M)$ and $F = D_i(F)$ so the result follows trivially. Suppose by induction that $D_j(F)$ is acyclic for all $j < i$ with i - j even. Let J_j be defined as above, i.e., $J_j = M - \cup X_\ell$. Note that J_i is the union of the top two stratum of M. By reversing the above argument we prove by another induction that

$F|_{J_j}$ is acyclic and, in particular, that $F|_{J_i}$ is acyclic. But J_i is a $SO(i)$-bundle over $D_i(M)$ and $F|_{J_i}$ is a bundle map covering $D_i(F)$. Hence, it follows from the Comparison Theorem (for spectral sequences) that $D_i(F)$ is acyclic. □

The remainder of this appendix is devoted to the proof of Lemma 1. The conclusion of this lemma asserts that the maps

1) $F|_{X_i}: X_i \to X_i'$

2) $(F|_{X_i}, F|_{\nabla X_i}): (X_i, \nabla X_i) \to (X_i', \nabla X_i')$

3) $F|_{\nabla X_i}: \nabla X_i \to \nabla X_i'$

are each acyclic. Notice that it follows from the exact sequence of the pair that if any two are acyclic, then so is the third. Thus, it suffices to show the first two are acyclic. Each of these maps is the union of two bundle maps. For example, $F|_{X_i} = F|_{N_i} \cup F|_{N_{i-1}}$ where $F|_{N_j}$ is a bundle map covering $f|_{B_j}: B_j \to B_j'$. We propose to compute the homology of the map* from the Mayer Vietoris sequence of the pair and the Serre spectral sequence of each fiber bundle in the decomposition. That is to say, the following two techniques will be used in our homology computation.

<u>Technique 1</u>: If

$$g: (X \cup Y, X, Y, X \cap Y) \longrightarrow (X' \cup Y', X', Y', X' \cap Y')$$

is a map of 4-tuples, then there is a Mayer-Vietoris sequence

---

*The homology of a map is equal to the homology of the mapping cone.

$$\longrightarrow H_*(g|_{X\cap Y}) \longrightarrow H_*(g|_X) + H_*(g|_Y) \longrightarrow H_*(g) \longrightarrow \ \dots .$$

Similarly, if $h\colon (X \cup Y, C \cup D) \to (X' \cup Y', C' \cup D')$ is a map which preserves all pieces and all their intersections, then there is a relative Mayer-Vietoris sequence

$$\longrightarrow H_*(h|_{X\cap Y}, h|_{C\cap D}) \longrightarrow H_*(h|_X, h|_C) + H_*(h|_Y, h|_D) \longrightarrow H_*(h|_{X\cup Y}, h|_{C\cup D}) \longrightarrow .$$

Here $C \subset X$ and $D \subset Y$.

Technique 2: Suppose that $(X,Y) \to C$ is a bundle pair with fiber pair (H,J), that $(X',Y') \to C'$ is another such bundle pair with the same fiber and that $g\colon (X,Y) \to (X',Y')$ is a bundle map covering $\bar{g}\colon C \to C'$. Then there is a spectral sequence with E^2-term

$$H_p(\bar{g}; H_q(H,J))$$

converging to $H_{p+q}(g, g|_Y)$.

The proof of Lemma 1 is now in three steps.

Step 1: $D_i(M)$ is a manifold with involution and as such it can be written as the union of two pieces; namely, a tubular neighborhood of the fixed point set and the complement of this tubular neighborhood. The fixed point set is B_{i-1}. Let ν denote a tubular neighborhood of B_{i-1} in $D_i(M)$ and let $\widetilde{B}_i = \overline{D_i(M) - \nu}$. Then $\widetilde{B}_i/Z/2 \cong B_i$. Also, let $\partial_{i-1}\widetilde{B}_i = \widetilde{B}_i \cap \nu$ be the unit sphere bundle associated to ν. To simplify notation, let $\widetilde{f} = D_i(F)|_{\widetilde{B}_i}$ and let $f_\nu = D_i(F)|_\nu$. Step 1 consists of applying the Mayer-Vietoris sequence to the maps

$$D_i(F) = \tilde{f} \cup f_\nu$$

$$F|_{X_i} = F|_{N_i} \cup F|_{N_{i-1}}$$

and

$$(F|_{X_i}, F|_{\nabla X_i}) = (F|_{N_i} \cup F|_{N_{i-1}}, F|_{P_i} \cup F|_{Q_i})$$

in order to deduce the following three lemmas.

<u>Lemma 2</u>: $D_i(F)$ <u>is acyclic if and only if</u>

$$H_*(\tilde{f}|_{\partial_{i-1}\tilde{B}_i}) \longrightarrow H_*(\tilde{f}) + H_*(f_\nu)$$

<u>is an isomorphism</u>.

<u>Lemma 3:</u> $F|_{X_i}$ <u>is acyclic if and only if</u>

$$H_*(F|_{Y_i}) \longmapsto H_*(F|_{N_i}) + H_*(F|_{N_{i-1}})$$

<u>is an isomorphism</u>.

<u>Lemma 4</u>: $(F|_{X_i}, F|_{\nabla X_i})$ <u>is acyclic if and only if</u>

$$H_*(F|_{Y_i}, F|_{\nabla Y_i}) \longrightarrow H_*(F|_{N_i}, F|_{P_i}) + H_*(F|_{N_{i-1}}, F|_{Q_i})$$

<u>is an isomorphism</u>.

<u>Step 2</u>: Notice that $N_i \to B_i$ is a bundle with fiber

$$Z_i = O(n) \times_{O(n-i)} D(n-i,k-i),$$

where $D(n-i,k-i)$ denotes the unit disk in $M(n-i,k-i)$. Y_i can be regarded as a bundle in two ways. First,

$$Y_i = N_i|_{\partial_{i-1}B_i}.$$

Y_i can also be regarded as a sub-bundle of the unit sphere bundle $\Sigma N_{i-1} \to B_{i-1}$. Thus, Y_i is a bundle over $\partial_{i-1}B_i$ and also a bundle over B_{i-1}. We have spectral sequences $E(F|_{Y_i})$, $E(F|_{N_i})$, $E(F|_{N_{i-1}})$ with E^2 terms

$$E^2_{p,q}(F|_{Y_i}) = H_p(f|_{\partial_{i-1}B_i};H_q(Z_i))$$

$$E^2_{p,q}(F|_{N_i}) = H_p(f|_{B_i};H_q(Z_i))$$

$$E^2_{p,q}(F|_{N_{i-1}}) = H_p(f|_{B_{i-1}};H_q(Z_{i-1})).$$

These converge to $H_{p+q}(F|_{Y_i})$, $H_{p+q}(F|_{N_i})$ and $H_{p+q}(F|N_{i-1})$, respectively. Here $f: B \to B'$ denotes the induced map of orbit spaces. Since the inclusion $Y_i \hookrightarrow N_i$ is a bundle map and $Y_i \hookrightarrow N_{i-1}$ is a sub-bundle, there is an induced map of spectral sequences

$$E(F|_{Y_i}) \longrightarrow E(F|_{N_i}) + E(F|_{N_{i-1}})$$

converging to

$$H_*(F|_{Y_i}) \longrightarrow H_*(F|_{N_i}) + H_*(F|_{N_{i-1}}).$$

Of course, the idea is to prove that if $D_i(F)$ is acyclic, then the map of spectral sequences is an isomorphism at the E^2-level and therefore, by the Comparison Theorem, that it converges to an isomorphism. Lemma 3 will then imply that $F|_{X_i}$ is acyclic. That the map is an isomorphism at the E^2-level will be proved from calculations of $H_*(Z_i)$, $H_*(Z_{i-1})$ and the map $H_*(Z_i) \to H_*(Z_{i-1})$.

<u>Remark 3</u>: At this point the reader's natural reaction should be that what we are trying to prove in Step 2 is false. Indeed, since Z_i is

homotopy equivalent to the Stiefel manifold $V_{n,i}$ of i-frames in n-space and since $H_*(V_{n,i}) \not\cong H_*(V_{n,i-1})$ it appears to be false that Lemma 2 implies that

$$H_p(f|_{\partial_{i-1}B_i};H_q(V_{n,i})) \cong H_p(f|_{B_i};H_q(V_{n,i})) + H_p(f|_{B_{i-1}};H_q(V_{n,i-1})).$$

However, we are saved by two facts:

1) The coefficients are twisted.

2) $H_*(f|_{B_i})$, $H_*(f|_{\partial_{i-1}B_i})$ and $H_*(f|_{B_{i-1}})$ are all odd torsion groups. Thus, the 2-torsion in the homology of the Stiefel manifolds plays no role.

<u>Step 3</u>: Next we verify the second part of Lemma 1, that is, we show that if $D_i(F)$ is acyclic then so is $(F|_{X_i}, F|_{\nabla X_i})$. The proof will be similar to Step 2, only now we must use the relative Mayer-Vietoris sequence of Lemma 4 and the spectral sequences $E(F|_{Y_i}, F|_{\nabla Y_i})$, $E(F|_{N_i}, F|_{P_i})$, $E(F|_{N_{i-1}}, F|_{Q_i})$ with E^2-terms:

$$E^2_{p,q}(F|_{Y_i}, F|_{\ Y_i}) = H_p(f|_{\partial_{i-1}B_i};H_q(Z_i,\partial Z_i))$$

$$E^2_{p,q}(F|_{N_i}, F|_{P_i}) = H_p(f|_{B_i};H_q(Z_i,\partial Z_i))$$

$$E^2_{p,q}(F|_{N_{i-1}}, F|_{Q_i}) = H_p(f|_{B_{i-1}};H_q(Z_{i-1},W_{i-1}).$$

Here W_{i-1} is the submanifold of ∂Z_{i-1} defined as

$$W_{i-1} = O(n) \times_{O(n-i+1)} U(n-i+1,k-i+1),$$

where $U(n,k)$ is the complement of a tubular neighborhood of the 1-stratum of $\partial D(n,k)$. Thus, to complete Step 3 we must also know something about $H_*(Z_i,\partial Z_i)$, $H_*(Z_{i-1},W_{i-1})$ and the map $H_*(Z_i,\partial Z_i) \to H_*(Z_{i-1},W_{i-1})$.

Now we proceed with the calculations necessary to complete Steps 2 and 3.

Lemma 5: Set $k' = k - i + 1$. Suppose that $D_i(F)$ is acyclic. We consider the action of $Z/2 = O(i)/SO(i)$ on $D_i(M)$.

1) If k' is even, then $Z/2$ acts trivially on $H_*(\tilde{f})$ and on $H_*(\tilde{f}|_{\partial_{i-1}\tilde{B}_i})$. (Here, $\tilde{f} = D_i(F)|_{\tilde{B}_i}$.)

2) If k' is odd, then $H_*(\tilde{f})^+ = 0$, also $H_*(\tilde{f}|_{\partial_{i-1}\tilde{B}_i})^+ \cong H_*(f|_{B_{i-1}})$ and $H_*(\tilde{f}|_{\partial_{i-1}\tilde{B}_i})^- \cong H_*(\tilde{f})$. Here $H_*(\ )^+$ is the subgroup on which $Z/2$ acts trivially and $H_*(\ )^-$ is the subgroup on which $Z/2$ acts via multiplication by -1.

Proof: Let $\nu \to B_{i-1}$ be the normal disk bundle of B_{i-1} in $D_i(M)$ and let $\Sigma\nu$ be the associated sphere bundle. The fiber is $D^{k'}$ and $Z/2$ acts on ν via the antipodal map on each fiber. This has degree $(-1)^{k'}$ on each fiber. Hence if k' is even, $Z/2$ acts trivially on $H_*(f_\nu,f_\nu|_{\Sigma\nu})$ and on $H_*(f_\nu|_{\Sigma\nu})$. By excision,

$$H_*(f_\nu,f_\nu|_{\Sigma\nu}) \cong H_*(D_i(F),\tilde{f})$$
$$\cong H_{*-1}(\tilde{f}).$$

Hence, $Z/2$ acts trivially on $H_*(\tilde{f})$. Since $f_\nu|_{\Sigma\nu} = \tilde{f}|_{\partial_{i-1}B_i}$, $Z/2$ is also trivial on $H_*(\tilde{f}|_{\partial_{i-1}B_i})$. This proves 1).

Let T be the nontrivial element of $Z/2$. If t is odd, then T_* is multiplication by -1 on

$$H_*(f_\nu, f_\nu|_{\Sigma\nu}) \cong H_{*-t}(f|_{B_{i-1}}) \otimes H_t(D^t, S^t).$$

As before it follows from excision and the sequence of the pair $(D_i(F), \tilde{f})$ that $H_*(f_\nu, f_\nu|_{\Sigma\nu}) \cong H_*(\tilde{f})$. Hence T_* is also multiplication by -1 on this group. Next consider the sequence of the pair $(D_i(F), f|_{B_{i-1}})$. Since $D_i(F)$ is acyclic, we conclude that

$$\begin{aligned} H_*(f|_{B_{i-1}}) &\cong H_{*+1}(D_i(F), f|_{B_{i-1}}) \\ &\cong H_{*+1}(\tilde{f}, \tilde{f}|_{\partial_{i-1}\tilde{B}_i}). \end{aligned}$$

Since B_{i-1} is the fixed point set, T_* acts trivially on $H_*(f|_{B_{i-1}})$; hence, it also acts trivially on $H_*(\tilde{f}, \tilde{f}|_{\partial_{i-1}\tilde{B}_i})$. Finally, by considering the sequence of $(\tilde{f}, \tilde{f}|_{\partial_{i-1}\tilde{B}_i})$, we conclude that

$$\begin{aligned} H_*(\tilde{f}|_{\partial_{i-1}\tilde{B}_i})^+ &\cong H_{*+1}(\tilde{f}, \tilde{f}|_{\partial_{i-1}\tilde{B}_i}) \\ &\cong H_*(f|_{B_{i-1}}) \end{aligned}$$

and that

$$H_*(\tilde{f}|_{\partial_{i-1}\tilde{B}_i})^- \cong H_*(\tilde{f}),$$

which proves 2). □

<u>Lemma 6</u>: <u>As before, let</u> $k' = k - i + 1$ <u>and suppose that</u> $D_i(F)$ <u>is acyclic. Let</u> Z^T <u>denote the system of twisted integer coefficients associated to the double covering</u> $\tilde{B}_i \to B_i$.

1) <u>If</u> k' <u>is even, then</u>

$$H_*(f|_{B_i}; Z^T) \cong 0 \cong H_*(f|_{\partial_{i-1}B_i}; Z^T)$$

$$H_*(f|_{B_i}; Z) \cong H_*(\tilde{f}; Z)$$

$$H_*(f|_{\partial_{i-1}B_i};Z) \cong H_*(\tilde{f}|_{\partial_{i-1}\tilde{B}_i};Z).$$

2) <u>If</u> k' <u>is odd, then</u>

$$H_*(f|_{B_i};Z^T) \cong H_*(\tilde{f};Z)$$

$$\cong H_*(f|_{\partial_{i-1}B_i};Z^T)$$

$$H_*(f|_{B_i};Z) = 0$$

$$H_*(f|_{\partial_{i-1}B_i};Z) \cong H_*(f|_{B_{i-1}};Z).$$

<u>Proof</u>: Let $\tilde{X} \to X$ be a double covering. There is an exact sequence

$$\longrightarrow H_*(X;Z/n) \longrightarrow H_*(\tilde{X};Z/n) \longrightarrow H_*(X;Z^T/n).$$

If n is odd, this sequence is short exact and

$$H_*(X;Z/n) \cong H_*(\tilde{X};Z/n)^+$$

$$H_*(X;Z^T/n) \cong H_*(\tilde{X};Z/n)^-.$$

If $\tilde{g}$: $\tilde{X} \to \tilde{X}'$ is a map of double covers, covering g: $X \to X'$, then

$$H_*(g;Z/n) \cong H_*(\tilde{g};Z/n)^+$$

and

$$H_*(g;Z^T/n) \cong H_*(\tilde{g};Z/n)^-,$$

whenever n is odd. If, in addition, $H_*(g;Z/2) = 0$ then these formulas imply that

$$H_*(g;Z) \cong H_*(\tilde{g};Z)^+$$

$$H_*(g;Z^T) \cong H_*(\tilde{g},Z)^-.$$

The lemma now follows from this observation and the calculations in Lemma 5. □

Remark 4: Suppose that $E \to B$ is a bundle with fiber F and structure group H. The E^2-term of the Serre spectral sequence is $E^2_{p,q} = H_p(B;H_q(F))$. The coefficients may be twisted. Thus we must keep track of the action of $\pi_1(B)$ on $H_*(F)$. Let us describe this action in terms of the structure group H. Let $H' \subset H$ be the kernel of the action of H on $H_*(F)$. Notice that H' contains the identity component of H. Let $\hat{E}$ be the associated principal H-bundle. Then $\hat{E}/H' \to B$ is a nontrivial regular covering associated to the homomorphism $\pi_1(B) \to H/H'$ given by the action of $\pi_1(B)$ on $H_*(F)$.

In our problem all the bundles involved are associated to normal orbit bundles (see Chapter III.A). Thus, for example, $N_i \to B_i$ and $(N_i,P_i) \to B_i$ have structure group $O(i) \times O(k-i)$ and are both associated to the principal bundle $E_i \to B_i$. Under the assumption $n - i$ is even, we will show that $SO(i) \times O(k-i)$ is the subgroup which acts ineffectively on $H_*(Z_i)$ and that is also the subgroup which acts ineffectively on $H_*(Z_i,\partial Z_i)$. Thus, in both cases the local coefficients are associated to the double covering $E_i/SO(i) \times O(k-i) = \tilde{B}_i \to B_i$.

In the case of the bundles $N_{i-1} \to B_{i-1}$ and $(N_{i-1},Q_i) \to B_{i-1}$, we will show, again under the assumption that $n - i$ is even, that the whole group $O(i-1) \times O(k-i+1)$ acts trivially on $H_*(Z_{i-1})$ and on $H_*(Z_{i-1},W_{i-1})$ and hence, that the coefficients are untwisted in these cases.

We now turn to the problem of computing the necessary information concerning the homology of the fibers.

Notation: Let $T_*(X)$ denote the torsion subgroup of $H_*(X)$ and let $F_*(X) = H_*(X)/T_*(X)$ be the free part.

Lemma 7: *Let* $V_{n,i}$ *be the Stiefel manifold of* i-*frames in* n-*space*. *Let* $p\colon V_{n,i} \to V_{n,i-1}$ *be the natural projection (which associates an* (i-1)-*frame to an* i-*frame by forgetting the last vector. Let* $Z/2 = O(i)/SO(i)$. *Then*

1) $T_*(V_{n,i})$ *is* 2-*torsion*.

2) *If* $n - i$ *is odd, then* $Z/2$ *acts trivially on* $H_*(V_{n,i})$.

3) *If* $n - i$ *is even, then* $p_*\colon F_*(V_{n,i})^+ \to F_*(V_{n,i-1})$ *is an isomorphism. Here* $F_*(V_{n,i})^+$ *denotes the subgroup fixed by* $Z/2$.

Proof: Statement 1) is well-known.

The map $p\colon V_{n,i} \to V_{n,i-1}$ is the projection map of a sphere bundle with fiber S^{n-i}. Without loss of generality we may represent the nontrivial element of $Z/2 = O(i)/SO(i)$ by

$$T = \begin{pmatrix} 1 & & & & 0 \\ & 1 & & & \\ & & \ddots & & \\ & & & 1 & \\ 0 & & & & -1 \end{pmatrix}$$

where T is the diagonal i by i matrix with 1's in every diagonal entry except the last which is -1. For this choice of T the following diagram commutes

$$\begin{array}{ccc} V_{n,i} & \xrightarrow{\cdot T} & V_{n,i} \\ & \searrow \quad \swarrow & \\ & V_{n,i-1} & \end{array}$$

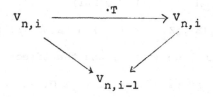

The action is the antipodal map of each fiber. Consider the Serre spectral sequence of the sphere bundle. The E^2-term is $H_p(V_{n,i-1};H_q(S^{n-i}))$. In n - i is odd, then T_* is trivial on $H_*(S^{n-i})$; hence, Z/2 acts trivially at the E^2-level and therefore, trivially on $H_*(V_{n,i})$. This proves 2).

If n - i is even, then T_* is multiplication by -1 on $H_{n-i}(S^{n-i})$. It follows that the Euler class of the sphere bundle has order two. Therefore,

$$F_*(V_{n,i}) \cong F_*(V_{n,i-1}) \otimes H_*(S^{n-i}).$$

Hence,

$$F_*(V_{n,i})^+ \cong F_*(V_{n,i-1}) \otimes H_0(S^{n-i})$$
$$\cong F_*(V_{n,i-1}),$$

which proves 3). □

<u>Lemma 8</u>: <u>Let</u> $(Z_i,\partial Z_i)$ <u>be as in Steps 2 and 3. Let</u> $H' \subset O(i) \times O(k-i)$ <u>be the subgroup which acts ineffectively on</u> $H_*(Z_i)$ <u>and let</u> $H'' \subset O(i) \times O(k-i)$ <u>be the subgroup that acts ineffectively on</u> $H_*(Z_i,\partial Z_i)$. <u>Then, if</u> n - i <u>is even,</u> $H' = H'' = SO(i) \times O(k-i)$.

<u>Proof</u>: Recall that $(Z_i,\partial Z_i) = O(n) \times_{O(n-i)} (D,\partial D)$ where $D = D(n-i,k-i)$. We see that Z_i is a vector bundle over $V_{n,i}$ and that $O(k-i)$ acts trivially on $V_{n,i}$. Hence, $SO(i) \times O(k-i) \subset H'$. On the other hand, by part 3) of the previous lemma, $O(i)/SO(i)$ acts nontrivially on $H_*(V_{n,i})$; therefore, $H' = SO(i) \times O(k-i)$.

Now consider $H_*(Z_i,\partial Z_i)$. $O(k-i)$ acts on $(Z_i,\partial Z_i)$ by multiplication (from the right) on $(D,\partial D)$, and the effect of the action on homotopy is independent of choice of $T' \in O(k-i) - SO(k-i)$. Let

$$T' = \begin{pmatrix} -1 & & & & & 0 \\ & 1 & & & & \\ & & 1 & & & \\ & & & \cdot & & \\ & & & & \cdot & \\ & & & & & \cdot \\ 0 & & & & & 1 \end{pmatrix} \in O(k-i).$$

Then T' acts by multiplying the first column of $D(n-i,k-i)$ by -1 and trivially on the other columns. Hence, T' is of degree $(-1)^{n-i} = 1$ on D. So, as before, $SO(i) \times O(k-i) \subset H''$. By the Thom Isomorphism Theorem, $H_*(Z_i, \partial Z_i) \cong H_*(V_{n,i}) \otimes H_*(D,\partial D)$. Since $O(i)/SO(i)$ acts non-trivially on $H_*(V_{n,i})$ it is also nontrivial on $H_*(Z_i, \partial Z_i)$. Hence $H'' = SO(i) \times O(k-i)$. □

<u>Lemma 9</u>: <u>Let</u> Z_{i-1} <u>and</u> W_{i-1} <u>be as in Steps 2 and 3. Then, if</u> $n - i$ <u>is even</u>, $O(i-1) \times O(k-i+1)$ <u>acts trivially on</u> $H_*(Z_{i-1})$ <u>and on</u> $F_*(Z_{i-1}, W_{i-1})$.

<u>Proof</u>: Since Z_{i-1} is a disk bundle over $V_{n,i-1}$, $O(i) \times O(k-i)$ acts trivially on $H_*(Z_i)$ by part 2) of Lemma 7. Recall that

$$W_{i-1} = O(n) \times_{O(n-i+1)} U(n-i+1,k-i+1)$$

where $U(n,k)$ is the complement of a tubular neighborhood of the 1-stratum in $\partial D(n,k)$. Thus, W_{i-1} is also a bundle over $V_{n,i-1}$ and therefore, by Lemma 7, $O(i-1)$ acts trivially on $H_*(W_{i-1})$. That $O(k-i+1)$ also acts trivially will follow from the spectral sequence and the next lemma which asserts that $O(k-i+1)$ acts trivially on $H_*(U(n-i+1,k-i+1))$/2-torsion. □

<u>Lemma 10</u>: <u>Let</u> $U(n,k)$ <u>be the complement of a tubular neighborhood of</u>

$\partial D(n,k)_1$ *in* $\partial D(n,k)$. *Then* $T_*(U(n,k))$ *is* 2-*torsion. If* n *is odd, then*

$$F_j(U(n,k)) \cong \begin{cases} Z & ; \quad j = 0, \ell-1 \\ 0 & ; \quad \text{otherwise} \end{cases}$$

where

$$\ell = \begin{cases} k(n-1) & ; \quad \text{if } k \text{ is even} \\ (k-1)(n-1) & ; \quad \text{if } k \text{ is odd} \end{cases}$$

Moreover, provided n *is odd*, $O(k)$ *acts trivially on* $F_*(U(n,k))$.

Proof: $\partial D(n,k)_1$ is a bundle over its image in the orbit space. The fiber is S^{n-1} and the structure group $O(1)$. The associated principal bundle can be identified with $\partial D(n,k)^{O(n-1)} = \partial D(1,k)$. Therefore

$$\partial D(n,k)_1 = S^{n-1} \times_{O(1)} S^{k-1}$$

where $O(1)$ acts via the antipodal map on both factors. Modulo 2-torsion, $H_*(\partial D(n,k)_1)$ is therefore just the invariant part of $H_*(S^{n-1} \times S^{m-1})$. In particular, this implies that $T_*(\partial D(n,k)_1)$ and $T_*(U(n,k))$ are both 2-torsion. Assuming that n is odd, $O(1)$ reverses the orientation of S^{n-1}. Therefore, if k is even, $H_*(S^{n-1} \times S^{k-1})^{O(1)}$ is nonzero only in dimensions 0 and $k-1$; while if k is odd, it is nonzero only in dimensions 0 and $n + k - 2$. Thus, $F_j(\partial D(n,k)_1) = Z$ precisely when $j = 0$ or $k - 1$ (if k is even) or when $j = 0$ or $n + k - 2$ (if k is odd). Since $U(n,k)$ is the complement of $\partial D(n,k)_1$ in the sphere $\partial D(n,k)$, it follows from Alexander duality that $F_j(U,n,k)) \cong Z$ precisely when $j = 0, \ell - 1$, where

$$\ell = \begin{cases} n(k-1) & ; \ k \ \text{even} \\ (n-1)(k-1) & ; \ k \ \text{odd}. \end{cases}$$

Let $v \in F_*(\ell D(n,k)_1)$ be a generator of the infinite cyclic group (in di ension $n + k - 2$ or $k - 1$) and let $u \in F_*(U(n,k))$ be its Alexander dual. Pick a reflection $T' \in O(k) - SO(k)$, say,

$$T' = \begin{pmatrix} -1 & & & & & & 0 \\ & 1 & & & & & \\ & & 1 & & & & \\ & & & \cdot & & & \\ & & & & \cdot & & \\ & & & & & \cdot & \\ 0 & & & & & & 1 \end{pmatrix}$$

Since n is odd $T'_*[\partial D(n,k)] = -[\partial D(n,k)]$. Next, notice that T' acts on $\partial D(n,k)_1 = S^{r-1} \times_{O(1)} S^{k-1}$ by a reflection on S^{k-1} and trivially on S^{n-1}. Therefore,

$$T'_* v = -v,$$

and hence, by duality,

$$T'_* u = u.$$

Thus, the action of $O(k)$ on $F_*(U(n,k))$ is trivial. □

<u>Lemma 11</u>: <u>Let</u> $j: (Z_i, \partial Z_i) \to (Z_{i-1}, W_{i-1})$ <u>be the inclusion of a fiber of the normal bundle of the</u> 1-<u>stratum of</u> ∂Z_{i-1}. <u>Suppose</u> $n - i$ <u>is even. Then the groups</u> $T_*(Z_i)$, $T_*(Z_i, \partial Z_i)$, $T_*(Z_{i-1})$ <u>and</u> $T_*(Z_{i-1}, W_{i-1})$ <u>are</u> 2-<u>torsion. Furthermore, the maps</u>

1) $j_*: F_*(Z_i)^+ \to F_*(Z_{i-1})$
2) $j_*: F_*(Z_i, \partial Z_i)^+ \to F_*(Z_{i-1}, W_{i-1})$

<u>are isomorphisms. Here</u> $F_*(\ )$ <u>is the free part of the homology and</u>

$F_*(Z_i)^+$ *and* $F_*(Z_i,\partial Z_i)^+$ *are the subgroups fixed by* $Z/2 = O(i)/SO(i)$.

Proof: The homology of a Stiefel manifold has no odd torsion. Since Z_i is a disk bundle over a Stiefel manifold and $(Z_i,\partial Z_i)$ is the Thom space of this disk-bundle, $T_*(Z_i)$ and $T_*(Z_i,\partial Z_i)$ are 2-torsion. By the previous lemma, if $n - i$ is even, then (Z_{i-1},W_{i-1}) also has the homology of the Thom space of a disk bundle over $V_{n,i-1}$ (at least, modulo 2-torsion). Hence, $T_*(Z_{i-1},W_{i-1})$ is also 2-torsion.

The following diagram clearly commutes

$$\begin{array}{ccc} (Z_i,\partial Z_i) & \xrightarrow{\;j\;} & (Z_{i-1},W_{i-1}) \\ \downarrow & & \downarrow \\ V_{n,i} & \xrightarrow{\;p\;} & V_{n,i-1}. \end{array}$$

Therefore, the fact that the first map $j_*: F_*(Z_i)^+ \to F_*(Z_{i-1})$ is an isomorphism, is immediate from part 3) of Lemma 7.

Next, set

$$k' = k - i + 1$$

$$n' = n - i + 1.$$

Recall that $(Z_i,\partial Z_i) = O(n) \times_{O(n-i)} (D,\partial D)$ where $D = D(n'-1,k'-1)$ is a disk of dimension $(n'-1)(k'-1)$.

Assertion:

$$F_*(Z_i,\partial Z_i)^+ \cong \begin{cases} F_*(V_{n,i})^+ \otimes H_*(D,\partial D) & ; \text{ if } k' \text{ is odd} \\ F_*(V_{n,i})^- \otimes H_*(D,\partial D) & ; \text{ if } k' \text{ is even.} \end{cases}$$

Here $F_*(V_{n,i})^-$ denotes the subgroup on which $Z/2$ acts by multiplication by -1. Let $[g,x]$ denote the image of (g,x) in $O(n) \times_{O(n-i)} D$.

Inclusion of the fiber $i: (D,S) \to (Z_i, \partial Z_i)$ is defined by $i(x) = [e,x]$. Let $T \in O(i) - SO(i)$ and let $h \in O(n-i)$ be a reflection. Regard T and h as elements in $O(n)$ via the standard inclusion $O(i) \times O(n-i) \subset O(n)$. Then $[T,x] = [Th, h^{-1}x] = Th[e, h^{-1}x]$ where $Th \in SO(n)$. Let y be the image of the fundamental class of $(D, \partial D)$ in $H_*(Z, \partial Z)$, i.e., $y = i_*[D] \in H_{(n'-1)(k'-1)}(Z_i, \partial Z_i)$. Since $SO(n)$ is connected, translation by Th is homotopic to the identity; consequently $T_*y = (Th)_*(h^{-1})_*y = (h^{-1})_*y$. Since h^{-1} is a reflection it has degree $(-1)^{k'-1}$ on D. Thus

$$T_*y = (-1)^{k'-1}y.$$

Since $F_*(Z_i, \partial Z_i)^+$ is the subgroup fixed by T_*, the Assertion follows.

Next, let

$$D' = D(n',k')$$

and

$$U = U(n',k') \subset \partial D'.$$

Let E be a tubular neighborhood of $\partial D_1'$ in $\partial D'$. The bundle $E \to \partial D_1'$ is $(-1)^{k'-1}$ orientable. By excision, $H_*(E, \partial E) \cong H_*(\partial D', U)$. Consider $F_*(D',U)$. This is nonzero only in dimension ℓ and in this dimension it is infinite cyclic, where, by Lemma 10,

$$\ell = \begin{cases} k'(n'-1) & ; \text{ if } k' \text{ is even} \\ (k'-1)(n'-1) & ; \text{ if } k' \text{ is odd.} \end{cases}$$

We now analyze the cases where k' is odd or even separately.

Suppose that <u>k' is odd</u>. Then $\ell = \dim D$ and $E \to \partial D_1'$ is an orientable bundle with fiber D. Consequently, the inclusions of the fiber $(D,\partial D) \hookrightarrow (Z_i,\partial Z_i)$ and $(D,\partial D) \hookrightarrow (E,\partial E)$ induce isomorphisms of infinite cyclic groups $H_\ell(D,\partial D) \cong H_\ell(Z_i,\partial Z_i)$ and $H_\ell(D,\partial D) \cong H_\ell(E,\partial E)$. Consequently, the inclusions $(D,\partial D) \subset (Z_i,\partial Z_i) \subset (E,\partial E) \subset (\partial D',U) \subset (D',U)$ induce isomorphisms $H_\ell(D,\partial D) \cong H_\ell(Z_i,\partial Z_i) \cong H_\ell(E,\partial E) \cong H_\ell(\partial D',U) \cong H_\ell(D',U) \cong \mathbb{Z}$. Therefore, by part 3) of Lemma 7,

$$\begin{aligned} F_*(Z_i,\partial Z_i)^+ &\cong F_*(V_{n,i})^+ \otimes H_*(D,\partial D) \\ &\cong F_*(V_{n,i-1}) \otimes F_*(D',U) \\ &\cong F_*(Z_{i-1},W_{i-1}) \end{aligned}$$

where the isomorphism are induced by the inclusion $(Z_i,\partial Z_i) \hookrightarrow (Z_{i-1},W_{i-1})$.

$$\begin{array}{ccc} (D,S) & & (D',U) \\ \downarrow & & \downarrow \\ (Z_i,\partial Z_i) & \hookrightarrow & (Z_{i-1},W_{i-1}) \\ \downarrow & & \downarrow \\ V_{n,i} & \xrightarrow{p} & V_{n,i-1} \end{array} .$$

Finally, suppose <u>k' is even</u>. In this case E is nonorientable and $\ell = k'(n'-1) = (n'-1) + \dim D$. Rather than regard $(Z_i,\partial Z_i)$ as a bundle over $V_{n,i}$, regard it as a bundle over $V_{n,i-1}$ with projection map $Z_i \to V_{n,i} \overset{p}{\to} V_{n,i-1}$ and with fiber $O(n') \times_{O(n'-1)} D$. Then

$$F_*(Z_i,\partial Z_i)^+ \cong F_*(V_{n,i-1}) \otimes F_*(O(n') \times_{O(n'-1)} (D,\partial D))^+ .$$

Also, since k' is even, $F_*(O(n') \times_{O(n'-1)} (D,\partial D))^+$ is non-zero only in dimension ℓ and

$$F_\ell(0(n') \times_{0(n'-1)} (D,\partial D))^+ \cong H_{n'-1}(S^{n'-1}) \otimes H_{(n'-1)(k'-1)}(D,\partial D)$$
$$\cong F_\ell(N,\partial N)$$
$$\cong F_\ell(D',U).$$

Thinking of $(Z_i,\partial Z_i) \subset (Z_{i-1},W_{i-1})$ as inclusion of a sub-bundle, we conclude that $F_*(Z_i,\partial Z_i)^+ \cong F_*(Z_{i-1},W_{i-1})$.

$$\begin{array}{ccc} 0(n') \times_{0(n'-1)} (D,\partial D) & \hookrightarrow & (D',U) \\ \downarrow & & \downarrow \\ (Z_i,\partial Z_i) & \hookrightarrow & (Z_{i-1},W_{i-1}) \\ & \searrow \quad \swarrow & \\ & V_{n,i-1} & \end{array}$$

□

<u>Proof of Lemma 1</u>: It remains to use Lemmas 6 and 11 to verify Steps 2 and 3. As usual $k' = k - i + 1$.

<u>Case 1, k' is even</u>: By Lemma 6, $H_*(f|_{B_i};Z^\tau) = 0$ and $H_*(f|_{\partial_{i-1}B_i};Z^\tau) = 0$. Consequently,

$$E^2_{p,q}(F|_{Y_i}) \cong H_p(f|_{\partial_{i-1}B_i};F_q(Z_i)^+)$$

$$E^2_{p,q}(F|_{N_i}) \cong H_p(f|_{B_i};F_q(Z_i)^+).$$

Here we are using the facts that $H_*(f|_{\partial_{i-1}B_i})$ and $H_*(f|_{B_i})$ are odd torsion and that $T_*(Z_i)$ is 2-torsion to replace $H_q(Z_i)^+$ by its free part $F_q(Z_i)^+$. Also, by Lemma 9, the coefficients of

$$E^2_{p,q}(F|_{N_{i-1}}) \cong H_p(f|_{B_{i-1}};F_q(Z_{i-1}))$$

are untwisted. By Lemmas 2 and 6, $H_p(f|_{\partial_{i-1}B_i}) \to H_p(f|_{B_i}) + H_p(f|_{B_{i-1}})$ is an isomorphism. Since by Lemma 11, $F_q(Z_i)^+ \to F_q(Z_{i-1})$ is an isomorphism, it follows that $E^2(F|_{Y_i}) \to E^2(F|_{N_i}) + E^2(F|_{N_{i-1}})$ is an isomorphism. Consequently $H_*(F|_{Y_i}) \to H_*(F|_{N_i}) + H_*(F|_{N_{i-1}})$ is an isomorphism and therefore $F|_{X_i}$ is acyclic by Lemma 2. That $(F|_{X_i}, F|_{\nabla X_i})$ is also acyclic is proved similarly.

<u>Case 2, k' is odd</u>: Using Lemma 6,

$$E^2_{p,q}(F|_{N_i}) = H_p(f|_{B_i}; F_q(Z_i)^-)$$

and $E^2_{p,q}(F|_{Y_i}) = C_+ \oplus C_-$, where

$$C_+ = H_p(f|_{\partial_{i-1}B_i}; F_q(Z_i)^+)$$

$$C_- = H_p(f|_{\partial_{i-1}B_i}; F_q(Z_i)^-).$$

By Lemma 6, $E^2_{p,q}(F|_{Y_i}) \to E^2_{p,q}(F|N_i)$ restricts to an isomorphism $C_- \cong E^2_{p,q}(F|_{N_i})$. By Lemmas 6 and 11, it restricts to an isomorphism $C_+ \cong E^2_{p,q}(F|_{N_{i-1}})$. Consequently $F|_{X_i}$ is acyclic in the case also. The proof that $(F|_{X_i}, F|_{\nabla X_i})$ is acyclic is similar. □

Finally, we should mention a proposition which is a corollary to Lemma 1.

<u>Proposition 12</u>: <u>Suppose that</u> $F: (M,\partial M) \to (M',\partial M')$ <u>is a stratified map of</u> k-<u>axial</u> O(n)-<u>manifolds. Fix an integer</u> i, <u>with</u> $0 \leq i \leq k+1$ <u>and</u> n − i <u>even. Further suppose that</u> $F|_{\partial M}$ <u>is acyclic and that for each</u> $j < i$ <u>with</u> n − j <u>even the map</u> $D_j(M) \to D_j(M')$ <u>is acyclic. Then the map</u> $\partial D_i(M) \to \partial D_i(M')$ <u>is acyclic</u>.

<u>Proof</u>: We have that $\partial D_i(M)$ is the union of $D_i(\partial M)$ and manifolds of the form $(\nabla X_j)/SO(i)$, where X_j is a regular neighborhood of the union of the j and (j-1) strata of $M^{O(n-i)}$. By the Main Lemma and Lemma 1,

the induced map is acyclic on each piece. Consequently, the proposition follows from Mayer-Vietoris sequences in the usual fashion.

□

This proposition is necessary to doing the inductive surgery in Chapter VI. Its use will be made more explicit in a joint paper with Hsiang, Morgan, in which the results of Chapter VI are proved.

Vol. 489: J. Bair and R. Fourneau, Etude Géométrique des Espaces Vectoriels. Une Introduction. VII, 185 pages. 1975.

Vol. 490: The Geometry of Metric and Linear Spaces. Proceedings 1974. Edited by L. M. Kelly. X, 244 pages. 1975.

Vol. 491: K. A. Broughan, Invariants for Real-Generated Uniform Topological and Algebraic Categories. X, 197 pages. 1975.

Vol. 492: Infinitary Logic: In Memoriam Carol Karp. Edited by D. W. Kueker. VI, 206 pages. 1975.

Vol. 493: F. W. Kamber and P. Tondeur, Foliated Bundles and Characteristic Classes. XIII, 208 pages. 1975.

Vol. 494: A Cornea and G. Licea. Order and Potential Resolvent Families of Kernels. IV, 154 pages. 1975.

Vol. 495: A. Kerber, Representations of Permutation Groups II. V, 175 pages. 1975.

Vol. 496: L. H. Hodgkin and V. P. Snaith, Topics in K-Theory. Two Independent Contributions. III, 294 pages. 1975.

Vol. 497: Analyse Harmonique sur les Groupes de Lie. Proceedings 1973–75. Edité par P. Eymard et al. VI, 710 pages. 1975.

Vol. 498: Model Theory and Algebra. A Memorial Tribute to Abraham Robinson. Edited by D. H. Saracino and V. B. Weispfenning. X, 463 pages. 1975.

Vol. 499: Logic Conference, Kiel 1974. Proceedings. Edited by G. H. Müller, A. Oberschelp, and K. Potthoff. V, 651 pages 1975.

Vol. 500: Proof Theory Symposion, Kiel 1974. Proceedings. Edited by J. Diller and G. H. Müller. VIII, 383 pages. 1975.

Vol. 501: Spline Functions, Karlsruhe 1975. Proceedings. Edited by K. Böhmer, G. Meinardus, and W. Schempp. VI, 421 pages. 1976.

Vol. 502: János Galambos, Representations of Real Numbers by Infinite Series. VI, 146 pages. 1976.

Vol. 503: Applications of Methods of Functional Analysis to Problems in Mechanics. Proceedings 1975. Edited by P. Germain and B. Nayroles. XIX, 531 pages. 1976.

Vol. 504: S. Lang and H. F. Trotter, Frobenius Distributions in GL_2-Extensions. III, 274 pages. 1976.

Vol. 505: Advances in Complex Function Theory. Proceedings 1973/74. Edited by W. E. Kirwan and L. Zalcman. VIII, 203 pages. 1976.

Vol. 506: Numerical Analysis, Dundee 1975. Proceedings. Edited by G. A. Watson. X, 201 pages. 1976.

Vol. 507: M. C. Reed, Abstract Non-Linear Wave Equations. VI, 128 pages. 1976.

Vol. 508: E. Seneta, Regularly Varying Functions. V, 112 pages. 1976.

Vol. 509: D. E. Blair, Contact Manifolds in Riemannian Geometry. VI, 146 pages. 1976.

Vol. 510: V. Poènaru, Singularités C^∞ en Présence de Symétrie. V, 174 pages. 1976.

Vol. 511: Séminaire de Probabilités X. Proceedings 1974/75. Edité par P. A. Meyer. VI, 593 pages. 1976.

Vol. 512: Spaces of Analytic Functions, Kristiansand, Norway 1975. Proceedings. Edited by O. B. Bekken, B. K. Øksendal, and A. Stray. VIII, 204 pages. 1976.

Vol. 513: R. B. Warfield, Jr. Nilpotent Groups. VIII, 115 pages. 1976.

Vol. 514: Séminaire Bourbaki vol. 1974/75. Exposés 453 – 470. IV, 276 pages. 1976.

Vol. 515: Bäcklund Transformations. Nashville, Tennessee 1974. Proceedings. Edited by R. M. Miura. VIII, 295 pages. 1976.

Vol. 516: M. L. Silverstein, Boundary Theory for Symmetric Markov Processes. XVI, 314 pages. 1976.

Vol. 517: S. Glasner, Proximal Flows. VIII, 153 pages. 1976.

Vol. 518: Séminaire de Théorie du Potentiel, Proceedings Paris 1972–1974. Edité par F. Hirsch et G. Mokobodzki. VI, 275 pages. 1976.

Vol. 519: J. Schmets, Espaces de Fonctions Continues. XII, 150 pages. 1976.

Vol. 520: R. H. Farrell, Techniques of Multivariate Calculation. X, 337 pages. 1976.

Vol. 521: G. Cherlin, Model Theoretic Algebra – Selected Topics. IV, 234 pages. 1976.

Vol. 522: C. O. Bloom and N. D. Kazarinoff, Short Wave Radiation Problems in Inhomogeneous Media: Asymptotic Solutions. V. 104 pages. 1976.

Vol. 523: S. A. Albeverio and R. J. Høegh-Krohn, Mathematical Theory of Feynman Path Integrals. IV, 139 pages. 1976.

Vol. 524: Séminaire Pierre Lelong (Analyse) Année 1974/75. Edité par P. Lelong. V, 222 pages. 1976.

Vol. 525: Structural Stability, the Theory of Catastrophes, and Applications in the Sciences. Proceedings 1975. Edited by P. Hilton. VI, 408 pages. 1976.

Vol. 526: Probability in Banach Spaces. Proceedings 1975. Edited by A. Beck. VI, 290 pages. 1976.

Vol. 527: M. Denker, Ch. Grillenberger, and K. Sigmund, Ergodic Theory on Compact Spaces. IV, 360 pages. 1976.

Vol. 528: J. E. Humphreys, Ordinary and Modular Representations of Chevalley Groups. III, 127 pages. 1976.

Vol. 529: J. Grandell, Doubly Stochastic Poisson Processes. X, 234 pages. 1976.

Vol. 530: S. S. Gelbart, Weil's Representation and the Spectrum of the Metaplectic Group. VII, 140 pages. 1976.

Vol. 531: Y.-C. Wong, The Topology of Uniform Convergence on Order-Bounded Sets. VI, 163 pages. 1976.

Vol. 532: Théorie Ergodique. Proceedings 1973/1974. Edité par J.-P. Conze and M. S. Keane. VIII, 227 pages. 1976.

Vol. 533: F. R. Cohen, T. J. Lada, and J. P. May, The Homology of Iterated Loop Spaces. IX, 490 pages. 1976.

Vol. 534: C. Preston, Random Fields. V, 200 pages. 1976.

Vol. 535: Singularités d'Applications Differentiables. Plans-sur-Bex. 1975. Edité par O. Burlet et F. Ronga. V, 253 pages. 1976.

Vol. 536: W. M. Schmidt, Equations over Finite Fields. An Elementary Approach. IX, 267 pages. 1976.

Vol. 537: Set Theory and Hierarchy Theory. Bierutowice, Poland 1975. A Memorial Tribute to Andrzej Mostowski. Edited by W. Marek, M. Srebrny and A. Zarach. XIII, 345 pages. 1976.

Vol. 538: G. Fischer, Complex Analytic Geometry. VII, 201 pages. 1976.

Vol. 539: A. Badrikian, J. F. C. Kingman et J. Kuelbs, Ecole d'Eté de Probabilités de Saint Flour V-1975. Edité par P.-L. Hennequin. IX, 314 pages. 1976.

Vol. 540: Categorical Topology, Proceedings 1975. Edited by E. Binz and H. Herrlich. XV, 719 pages. 1976.

Vol. 541: Measure Theory, Oberwolfach 1975. Proceedings. Edited by A. Bellow and D. Kölzow. XIV, 430 pages. 1976.

Vol. 542: D. A. Edwards and H. M. Hastings, Čech and Steenrod Homotopy Theories with Applications to Geometric Topology. VII, 296 pages. 1976.

Vol. 543: Nonlinear Operators and the Calculus of Variations, Bruxelles 1975. Edited by J. P. Gossez, E. J. Lami Dozo, J. Mawhin, and L. Waelbroeck, VII, 237 pages. 1976.

Vol. 544: Robert P. Langlands, On the Functional Equations Satisfied by Eisenstein Series. VII, 337 pages. 1976.

Vol. 545: Noncommutative Ring Theory. Kent State 1975. Edited by J. H. Cozzens and F. L. Sandomierski. V, 212 pages. 1976.

Vol. 546: K. Mahler, Lectures on Transcendental Numbers. Edited and Completed by B. Diviš and W. J. Le Veque. XXI, 254 pages. 1976.

Vol. 547: A. Mukherjea and N. A. Tserpes, Measures on Topological Semigroups: Convolution Products and Random Walks. V, 197 pages. 1976.

Vol. 548: D. A. Hejhal, The Selberg Trace Formula for PSL (2, ℝ). Volume I. VI, 516 pages. 1976.

Vol. 549: Brauer Groups, Evanston 1975. Proceedings. Edited by D. Zelinsky. V, 187 pages. 1976.

Vol. 550: Proceedings of the Third Japan – USSR Symposium on Probability Theory. Edited by G. Maruyama and J. V. Prokhorov. VI, 722 pages. 1976.

Vol. 551: Algebraic K-Theory, Evanston 1976. Proceedings. Edited by M. R. Stein. XI, 409 pages. 1976.

Vol. 552: C. G. Gibson, K. Wirthmüller, A. A. du Plessis and E. J. N. Looijenga. Topological Stability of Smooth Mappings. V, 155 pages. 1976.

Vol. 553: M. Petrich, Categories of Algebraic Systems. Vector and Projective Spaces, Semigroups, Rings and Lattices. VIII, 217 pages. 1976.

Vol. 554: J. D. H. Smith, Mal'cev Varieties. VIII, 158 pages. 1976.

Vol. 555: M. Ishida, The Genus Fields of Algebraic Number Fields. VII, 116 pages. 1976.

Vol. 556: Approximation Theory. Bonn 1976. Proceedings. Edited by R. Schaback and K. Scherer. VII, 466 pages. 1976.

Vol. 557: W. Iberkleid and T. Petrie, Smooth S^1 Manifolds. III, 163 pages. 1976.

Vol. 558: B. Weisfeiler, On Construction and Identification of Graphs. XIV, 237 pages. 1976.

Vol. 559: J.-P. Caubet, Le Mouvement Brownien Relativiste. IX, 212 pages. 1976.

Vol. 560: Combinatorial Mathematics, IV, Proceedings 1975. Edited by L. R. A. Casse and W. D. Wallis. VII, 249 pages. 1976.

Vol. 561: Function Theoretic Methods for Partial Differential Equations. Darmstadt 1976. Proceedings. Edited by V. E. Meister, N. Weck and W. L. Wendland. XVIII, 520 pages. 1976.

Vol. 562: R. W. Goodman, Nilpotent Lie Groups: Structure and Applications to Analysis. X, 210 pages. 1976.

Vol. 563: Séminaire de Théorie du Potentiel. Paris, No. 2. Proceedings 1975-1976. Edited by F. Hirsch and G. Mokobodzki. VI, 292 pages. 1976.

Vol. 564: Ordinary and Partial Differential Equations, Dundee 1976. Proceedings. Edited by W. N. Everitt and B. D. Sleeman. XVIII, 551 pages. 1976.

Vol. 565: Turbulence and Navier Stokes Equations. Proceedings 1975. Edited by R. Temam. IX, 194 pages. 1976.

Vol. 566: Empirical Distributions and Processes. Oberwolfach 1976. Proceedings. Edited by P. Gaenssler and P. Révész. VII, 146 pages. 1976.

Vol. 567: Séminaire Bourbaki vol. 1975/76. Exposés 471-488. IV, 303 pages. 1977.

Vol. 568: R. E. Gaines and J. L. Mawhin, Coincidence Degree, and Nonlinear Differential Equations. V, 262 pages. 1977.

Vol. 569: Cohomologie Etale SGA 4½. Séminaire de Géométrie Algébrique du Bois-Marie. Edité par P. Deligne. V, 312 pages. 1977.

Vol. 570: Differential Geometrical Methods in Mathematical Physics, Bonn 1975. Proceedings. Edited by K. Bleuler and A. Reetz. VIII, 576 pages. 1977.

Vol. 571: Constructive Theory of Functions of Several Variables, Oberwolfach 1976. Proceedings. Edited by W. Schempp and K. Zeller. VI, 290 pages. 1977

Vol. 572: Sparse Matrix Techniques, Copenhagen 1976. Edited by V. A. Barker. V, 184 pages. 1977.

Vol. 573: Group Theory, Canberra 1975. Proceedings. Edited by R. A. Bryce, J. Cossey and M. F. Newman. VII, 146 pages. 1977.

Vol. 574: J. Moldestad, Computations in Higher Types. IV, 203 pages. 1977.

Vol. 575: K-Theory and Operator Algebras, Athens, Georgia 1975. Edited by B. B. Morrel and I. M. Singer. VI, 191 pages. 1977.

Vol. 576: V. S. Varadarajan, Harmonic Analysis on Real Reductive Groups. VI, 521 pages. 1977.

Vol. 577: J. P. May, E_∞ Ring Spaces and E_∞ Ring Spectra. IV, 268 pages. 1977.

Vol. 578: Séminaire Pierre Lelong (Analyse) Année 1975/76. Edité par P. Lelong. VI, 327 pages. 1977.

Vol. 579: Combinatoire et Représentation du Groupe Symétrique, Strasbourg 1976. Proceedings 1976. Edité par D. Foata. IV, 339 pages. 1977.

Vol. 580: C. Castaing and M. Valadier, Convex Analysis and Measurable Multifunctions. VIII, 278 pages. 1977.

Vol. 581: Séminaire de Probabilités XI, Université de Strasbourg. Proceedings 1975/1976. Edité par C. Dellacherie, P. A. Meyer et M. Weil. VI, 574 pages. 1977.

Vol. 582: J. M. G. Fell, Induced Representations and Banach *-Algebraic Bundles. IV, 349 pages. 1977.

Vol. 583: W. Hirsch, C. C. Pugh and M. Shub, Invariant Manifolds. IV, 149 pages. 1977.

Vol. 584: C. Brezinski, Accélération de la Convergence en Analyse Numérique. IV, 313 pages. 1977.

Vol. 585: T. A. Springer, Invariant Theory. VI, 112 pages. 1977.

Vol. 586: Séminaire d'Algèbre Paul Dubreil, Paris 1975-1976 (29ème Année). Edited by M. P. Malliavin. VI, 188 pages. 1977.

Vol. 587: Non-Commutative Harmonic Analysis. Proceedings 1976. Edited by J. Carmona and M. Vergne. IV, 240 pages. 1977.

Vol. 588: P. Molino, Théorie des G-Structures: Le Problème d'Equivalence. VI, 163 pages. 1977.

Vol. 589: Cohomologie l-adique et Fonctions L. Séminaire de Géométrie Algébrique du Bois-Marie 1965-66, SGA 5. Edité par L. Illusie. XII, 484 pages. 1977.

Vol. 590: H. Matsumoto, Analyse Harmonique dans les Systèmes de Tits Bornologiques de Type Affine. IV, 219 pages. 1977.

Vol. 591: G. A. Anderson, Surgery with Coefficients. VIII, 157 pages. 1977.

Vol. 592: D. Voigt, Induzierte Darstellungen in der Theorie der endlichen, algebraischen Gruppen. V, 413 Seiten. 1977.

Vol. 593: K. Barbey and H. König, Abstract Analytic Function Theory and Hardy Algebras. VIII, 260 pages. 1977.

Vol. 594: Singular Perturbations and Boundary Layer Theory, Lyon 1976. Edited by C. M. Brauner, B. Gay, and J. Mathieu. VIII, 53[illegible] pages. 1977.

Vol. 595: W. Hazod, Stetige Faltungshalbgruppen von Wahrscheinlichkeitsmaßen und erzeugende Distributionen. XIII, 157 Seiten. 197[illegible]

Vol. 596: K. Deimling, Ordinary Differential Equations in Banach Spaces. VI, 137 pages. 1977.

Vol. 597: Geometry and Topology, Rio de Janeiro, July 1976. Proceedings. Edited by J. Palis and M. do Carmo. VI, 866 pages. 197[illegible]

Vol. 598: J. Hoffmann-Jørgensen, T. M. Liggett et J. Neveu, Ecole d'Eté de Probabilités de Saint-Flour VI - 1976. Edité par P.-L. Hennequin. XII, 447 pages. 1977.

Vol. 599: Complex Analysis, Kentucky 1976. Proceedings. Edited by J. D. Buckholtz and T. J. Suffridge. X, 159 pages. 1977.

Vol. 600: W. Stoll, Value Distribution on Parabolic Spaces. V[illegible] 216 pages. 1977.

Vol. 601: Modular Functions of one Variable V, Bonn 1976. Proceedings. Edited by J.-P. Serre and D. B. Zagier. VI, 294 pages. 1977.

Vol. 602: J. P. Brezin, Harmonic Analysis on Compact Solvmanifolds. VIII, 179 pages. 1977.

Vol. 603: B. Moishezon, Complex Surfaces and Connected Sums of Complex Projective Planes. IV, 234 pages. 1977.

Vol. 604: Banach Spaces of Analytic Functions, Kent, Ohio 197[illegible] Proceedings. Edited by J. Baker, C. Cleaver and Joseph Diestel. [illegible] 141 pages. 1977.

Vol. 605: Sario et al., Classification Theory of Riemannian Manifolds. XX, 498 pages. 1977.

Vol. 606: Mathematical Aspects of Finite Element Methods. Proceedings 1975. Edited by I. Galligani and E. Magenes. VI, 362 pages. 1977.

Vol. 607: M. Métivier, Reelle und Vektorwertige Quasimartingale und die Theorie der Stochastischen Integration. X, 310 Seiten. 197[illegible]

Vol. 608: Bigard et al., Groupes et Anneaux Réticulés. XIV, 3[illegible] pages. 1977.